Social Geographies

Ruth Panelli

Social Geographies
From Difference to Action

SAGE Publications
London • Thousand Oaks • New Delhi

First published 2004

SAGE Publications Ltd
6 Bonhill Street
London EC2A 4PU

SAGE Publications Inc.
2455 Teller Road
Thousand Oaks, California 91320

SAGE Publications India Pvt Ltd
B-42, Panchsheel Enclave
Post Box 4109
New Delhi 100 017

British Library Cataloguing in Publication data

A catalogue record for this book is available from the British Library

ISBN 0 7619 6893 8
ISBN 0 7619 6894 6 (pbk)

Library of Congress Control Number 2003103975

Typeset by C&M Digitals (P) Ltd., Chennai, India
Printed in Great Britain by The Cromwell Press Ltd, Trowbridge, Wiltshire

In memory of my father, Alexander Panelli, who taught me the
fascination and confidence to always 'have a theory' about things –
no matter how big or small.

&

For Maija and Tālis, with love, and in the hope that continued critical
social thought will contribute to a fairer and more compassionate world
for you and yours.

Contents

List of Boxes

List of Figures

List of Tables

Preface

How do we describe ourselves?
Where have we, do we, will we, live our lives?
Why are differences between people a source of tension?
How can social change occur?

Social geography can assist in addressing these questions. It provides ways of understanding and living in our contemporary world. This book gives one account of how geographers have studied 'western' societies and provided concepts, theories and arguments about human life and interaction in varying contexts defined via places, power relations, and systems of economic and cultural relations.

One of the core foci of social geography is the recognition and critique of social difference and the power relations this involves. In the (too many) years it has taken to complete this manuscript, questions of difference and power have increased as key academic and international concerns. Personally, I have worked in a country that continues to struggle with issues surrounding bicultural and multicultural tensions; I have thought a good deal about when and how I sometimes claim to be an 'Australian'; and I have experienced my own and others' responses to the events of 11 September 2001 and the resulting polarization of ideas and attitudes about social differences based on ethnicity and religion. Walking in anti-war marches has become a crucial social action again as new generations of people think through the political and humanitarian implications associated with social differences and the way social differences can be mobilized for political, economic and military ends. While the subject of this book is not focused on international politics or the 'war on terrorism', critiques of the underlying ways social differences are constructed and the power relations involved are as relevant now as they ever were.

So in writing this book I have assembled an account of how geographers have constructed successive literatures on the differences that separate us, and the ways in which we may live and think across and beyond those differences as we negotiate questions of identity, power and social action. While mapping patterns of social difference has long been an activity in social geography, so to has been the socio-political drive to critique uneven power relations and the dominant constructions of difference and identity that leave little space for heterogeneous (including alternative) social forms. Thus, the structure of this book has been organized to outline the various theoretical approaches to social geography, and the core literatures on class, gender, ethnicity and sexuality. But it also seeks to recognize that differences are only part of the story and that what we do because of, and beyond, these differences is equally important. Consequently, this book culminates with a discussion of contemporary critique and hope as they occur in investigations of the politics of identity, the operation of power relations, and the potential and possibilities within various forms of social action.

This account does not attempt to be an encyclopaedic presentation of social geography. Instead, it is written as a companion to what I believe should be the most important study strategy for undergraduate students – the act of reading and critically engaging with theoretical questions and research being constructed in original journal articles and books. If this book provides a resource and support for this process, it will have achieved one of its major objectives.

A second objective for the book lies in the desire to draw together the ways social geographic knowledge can be developed from many voices. Consequently, I have included material from various academic sources, from personal accounts of individual geographers' practice (see Chapter 2 and Part V: Individual Social Geographies), and from a range of student reflections and observations. I have also sought to include a diverse selection of examples from within the scope of Anglo-American and Antipodean geographies, as outlined in Chapter 1. This involves material from urban and rural settings in Australia, Canada, New Zealand, the United Kingdom and the United States of America.

In writing this book, I have made a number of decisions that reflect my academic base, and which should be explained. In terms of conventions, I use the word Maori (rather than Maaori, Māori, etc.) since a variety of debates surround the preferred usage and spelling of this word and differences exist across *iwi* (tribes) and in different settings and time periods. Likewise, the term Aotearoa is not linked with, or equated with, the word New Zealand (as in Aotearoa/New Zealand) since different *iwi* understand the term to refer to one or more of the islands making up this country. There is no uniform recognition across all *iwi* that the term refers to all the islands within this country.

For ease of reading, a number of other standard treatments have been employed. First, a variety of non-English words have been placed in italics. Second, a series of terms, which are defined and discussed in more detail in the glossary, are placed in bold italics – these include theoretical terms and concepts as well as key Maori words. Third, a summary of key points is provided at the end of Chapters 2–10, to distil and highlight the central issues and contentions being made at each stage. Finally, while numerous references are cited in most chapters in order to capture some of the diversity of geographies being written on any one topic, a list of useful texts is also given at the end Chapters 2–10 in order to encourage and direct starting points for further reading. It is hoped that the combined reading of this book and the suggested references will enhance understanding and respect for social differences and the possibilities of critiquing and reconstructing unequal material conditions and power relations.

Acknowledgements

In writing this book, I first need to thank those students who have taken Social Geography papers with me over the years. You are the people who first inspired me to think carefully about my own perspectives on the purpose, character and possibilities of social geography. Your willingness to enrol in social geography papers has continued to support my passion for advocating this type of geography as a way to approach the social world with critical thought, care and passion.

Next, this book owes its origin to the interest of and kind invitation from Robert Rojek of Sage Publications, a source of endless patience and encouragement. It is also a great pleasure to record that the book would never have happened without the crucial, constant and generous encouragement and support of Richard Welch – there is no single way to thank you, but this is one first step.

I am also grateful for important advice at crucial moments from Nicola Peart; and for essential practical assistance that thas been provided by Jessica Hattersley and Anna Kraack (making careful and cheerful reading and reference chasing), and Bill Mooney (drafting, editing, scanning and troubleshooting for numerous figures with great patience and care); and for various advice and guidance kindly given by Richard Bedford, Paul Cloke, Robin Kearns, Thelma Fisher and Brian Roper.

I also greatly appreciate the support provided by staff of the geography departments at the University of Bristol, University of Glasgow and University of Otago, where substantial parts of this book were drafted.

Special thanks is also due to individual geographers who generously participated in interviews and reflections about their work. In order of appearance they are: Paul Cloke, Geraldine Pratt, Andrew Herod, Jo Little, Kay Anderson, Larry Knopp, Karen M. Morin, Michael Woods, Katherine Gibson and Julie Graham. Further thanks are due to Louise Johnson, Claire Freeman and Paul Routledge for their generosity in sharing their photographs (Box 4.1, and Figures 8.1, 9.1 and 9.7 respectively).

The author and publishers wish to thank the following for permission to reproduce material:

Box 5.1, Pearson Education, for Duce (1956), p. 78;
Box 5.3, Routledge, for Metge (1967), pp. 173, 175, 176, 177, 178–9 and Fig. 5;
Boxes 7.1 and 8.1, Judy Horacek, for cartoons;
Box 8.2, *Australian Farm Journal*, photograph (1994), p. 37;
Figure 2.1, Hodder Arnold, for Harvey (1969), Figs 4.2 and 4.3;
Figure 3.3, Verso, for Wright (1985), Table 3.3;
Figure 4.1, *The Dominion Post*;
Figure 5.1, American Geographical Society (NY), for Morrill (1965), Figs 11–12;
Figure 5.2, Blackwell Publishing, for Gilbert (1998), Fig. 2;
Figure 5.3. Pearson Education Australia, for Pawson (1992), Fig. 2.1;
Figure 6.1, Blackwell Publishing, for Symanski (1974), Figs 6–7;
Figure 9.2, Allied Press, *Otago Daily Times*, for photograph; and
Figure 9.5, Cambridge University Press, for Melucci (1996), Figs 1–2.

Individuals who have made crucial readings of various chapters and key points of this project are greatly appreciated: Peter Jackson, Jo Little, Richard Welch, Paul Cloke, Geraldine Pratt and Larry Knopp. Sincere appreciation is also extended to colleagues, postgraduates and teaching assistants who have variously encouraged and supported me, or ignored my occasional bad humour and pre-occupation with the teaching and writing of this social geography, specifically: Jo, Anna, Karen, Nicola, Claire, Sarah, Deirdre, Margot, Jen, Murray and Zannah. Special thanks also to Deirdre, for initiative with early sketches for Box 3.4.

Finally, to those special people who have walked beside me (sometimes very generously putting up with me) throughout this project and the other momentous events that have coincided with it – my love and thanks go especially to you: Richard, Maija, Tālis, Dorothy, Alex, Anna, Eleanor, Jo and Kirsty.

INTRODUCTIONS AND NEGOTIATIONS

PART I

Social geography is a body of knowledge and a set of practices by which scholars look at, and seek to understand, the social world. It is a strikingly diverse sub-discipline of human geography that has many overlapping interests with other forms of geography rather than any fixed or strict boundaries. Its diversity occurs since the topics and approaches undertaken have varied over time and according to individual geographers' interests and politics. Consequently, *Introducing Social Geographies: From Difference to Action* presents a cameo of that diversity and Part I opens with two chapters that introduce and contextualize the sub-discipline.

Chapter 1 starts by sketching out the types of social relations and places that fascinate social geographers before explaining that (as with other academic knowledge) social geography is an explicitly constructed field of knowledge. This means scholars frequently concentrate on commonly agreed topics – especially the differences and relations between people and the places and spaces they use and shape in creating their lives. It also means social geographers cluster into groups of scholars who *practise* (pursue, design, construct, promote, and even fight for) certain ways of constructing their social geography. These preferences depend on the context and culture in which they are working (e.g. a British department enthusiastic about the methods and powers of ***spatial science***[1] perspectives in the 1970s, or a Canadian department focused on ***humanist*** approaches in the 1980s, or an Australian department energized by ***poststructural*** and ***postcolonial*** debates and challenges in the 1990s). These contexts and preferences result in individual scholars taking up positions and 'writing … from somewhere' (section 1.2), from locations that are physical, cultural, political and ***epistemological***.

A more detailed account of these different approaches to social geography is presented in Chapter 2. This chapter commences with questions about how we can devise projects and knowledge about the social world. It explains how scientific approaches to knowledge have been incorporated in social geography before turning to an overview of the philosophical and theoretical perspectives that have been adopted by different groups of scholars over time. While readers

---

1. As noted in the Preface, all terms appearing in **bold** *italics* are included in the Glossary.

may be tempted to skip this chapter, it is a core foundation for the rest of the book and provides a basis for understanding why geographers have differed in their approaches to topics, e.g. ethnicity, gender, identity and so forth. The chapter also provides two detailed sketches of how individual geographers have tackled and worked their way through the theoretical trends and approaches that have swept across social geography at different times. The chapter closes with further emphasis on the *positionality* of our thinking and writing about social geography. This provides encouragement in making self-critical reflections and explicit decisions about how we view people, social experiences and the real-life (and academic tensions) that result.

1 Contemporary Social Geographies: Perspectives on Difference, Identity, Power and Action

Social geography provokes thought and challenges us with the possibilities and opportunities it provides. It gives us the chance to ask questions, construct explanations – and discover yet more questions – about where and how social differences and interaction occur. Gibson-Graham's analysis of life in Australian mining towns illustrates some of these possibilities (see Box 1.1). Some social relations between men, women and children appear similar. However, comparisons within towns and states (Queensland, vs New South Wales) show that individual households and groups of women experience mining life – and its social costs – in very different ways.

More generally, social geography inquiries may build from an awareness of contrasting social lives in different suburbs or rural towns; or it may develop from recognizing different behaviours and interaction occurring in a public square, workplace or pub. For me, social geography became relevant after I completed a fourth-year economic geography dissertation and was left with questions about how restructuring in a manufacturing industry affected workers and their families differently in various Australian cities. Gibson-Graham's (1996a) work, summarized in Box 1.1, raises parallel issues, showing how restructured work conditions affected personal, parental and community relations. This type of geography shows how individual, family and community life is closely linked. Gibson-Graham explores contrasting theories to account for the diverse ways men and women negotiate their economic and social environment, and in some cases, become politically active because of their situations. This initial example emphasizes the fact that social geography can be established wherever there is a variety of people relating in diverse ways and acting to organize their lives in both a material and socio-cultural sense. In these settings, attention to social difference and interaction is usually highlighted as geographers acknowledge that these social differences occur unevenly over space and through the constructions of (and even struggles within) specific places (e.g. in homes, on streets, in workplaces and so forth) (Valentine, 2001).

Box 1.1 Social conditions and implications of life in Australian coal-mining towns

Katherine Gibson is a geographer who has studied Australian coal-mining for many years. Together with Julie Graham, she investigated both the economic and social processes and changes that have shaped the lives of men, women and children in such industries. In recent years they have written collaboratively as J.K. Gibson-Graham. Excerpts of this work give a useful summary of the social conditions and challenges that these mining communities face. They also illustrate the types of issue social geography often concentrates on. They write:

> Men and women in [Australian] mining towns see themselves as engaged in a joint project ... a good upbringing for their children, a house of their own on the coast, a comfortable retirement or a different life, ... after savings have been accumulated. Commitment to this joint project (and the feudal domestic class process[1] it promotes) is sustained by a discourse of love and companionship between partners. (1996a: 228)

> In the late 1980s the coal industry entered a crisis. ... With employment levels declining the Combined Mining Unions were forced ... to accept the ruling of the Coal Industry Tribunal ... and new work practices were instituted in 1988. As part of the move towards greater 'flexibility' (for the companies), the new award involved widespread adoption of a new work roster called the seven-day roster. (1996a: 225)[2]

> The effect of this decision on the community and ... upon women has been great. For many, the domestic work of women has risen, and the established companionship patterns of mothers, fathers and children have largely been destroyed. (1996a: 226)

> The weekdays off between shifts and the long break between roster cycles allow men and their nonworking wives to see each other – but at times that do not coincide with children's or other friends' time off. ... The seven-day roster has stripped away the activities and notions of the family. (1996a: 228)

> In Central Queensland miners' wives organized no public opposition to the seven-day roster. By contrast in the Hunter Valley, an older established coal-mining region in New South Wales, women successfully organized opposition to its institution on the grounds of its incompatibility with family and community life. (1996a: 230)

1 A 'feudal' domestic class refers to arrangements where one partner (usually a woman) contributes labour as a wife and mother that is appropriated by the other partner (usually a man) in return for provision of shelter and social position associated with their conditions as a miner (e.g. access to housing and services).

2 The seven-day roster involved eight-hour shifts on seven consecutive days, afternoons or nights. After seven shifts, workers had one–two days off and at the end of a block of three seven-day shifts they received four days off from work.

Hamnett (1996: 3) defines social geography as 'the geography of social structures, social activities and social groups across a wide range of human societies'. Yet it is perhaps more complicated – and more exciting – than that, for social geographers are prepared to investigate the intimate connections and collections of interactions that occur between diverse people and the spaces and places in which this occurs. For instance, with Gibson-Graham, we can ask why some men and women in Australian coal-mining towns faced domestic struggles and divorce, while others worked out patterns of family and social life that absorbed (even accepted) the industry changes. We can also ask why NSW miners' wives became politically active against the work rosters, while Queensland miners' wives did not. Questions of this kind invite us to consider issues of gender, class, identity and political agency – some of the concepts that structure the remainder of this book.

Thus, social geography can be thought of as a focused curiosity and an explicit act of constructing (researching, mapping, writing) geographies that:

- recognize forms of social difference and interaction; and
- acknowledge that these differences occur unevenly over space and through the construction of (and even struggles within) specific places.

This type of social geography involves us in choosing appropriate theories and research practices in order to investigate and write work that respects difference and highlights uneven patterns and struggles. These matters are discussed in Chapter 2 since contemporary social geographies have developed from a diverse heritage of theoretical and empiric histories. The chapter will show that debates and tensions arising from this diversity provide a complex but stimulating environment for current work. Coming to terms with these debates is an important step in recognizing that social geographers frequently wish to do more than record, organize and (re)present social differences and interactions. They dare to ask why they occur. For instance, Gibson-Graham (1996a) sought reasons for the enormous hardship being faced by mining families and communities. By working alongside some of these people, Katherine Gibson (1993) also recognized the need to acknowledge and account for the very different relations and choices she found in contrasting coal-mining regions.

In general, by asking why differences occur, social geographers must consider different research perspectives or forms of explanation in order to select and address their questions – and (re)present the answers they construct. Explicitly or not, they position themselves, their practice and their writing in different ways, which in turn are both personal and political acts. These issues are explained further in Chapter 2 as it addresses some of the circumstances surrounding the construction of different geographic knowledges. Social geography is shown to be a creative inquiry that (implicitly or explicitly) negotiates both scientific and political issues associated with establishing academic geographic knowledge.

1.1 BOOK STRUCTURE

In writing this book I have not attempted to produce a comprehensive description of all forms of social geography. Other references and edited collections tackle this job in various ways (see for example Hamnett, 1996; Jackson and Smith, 1984; Pain et al., 2001; Valentine, 2001). Instead, *Social Geographies: From Difference of Action* presents a specific commentary on social geography as it appears primarily in Anglo-American and Antipodean contexts. The book confines itself to these contexts for two reasons. First, Anglo-American and Antipodean geographies have shared a common, dominating socio-economic system – capitalism. This has influenced the development of different types of geographic theory and the recognition of different research subjects common to the capitalist societies in these countries. Second, social geographies being written beyond Anglo-American and Antipodean contexts are more often considered within studies of (economic) development and (postcolonial) political and cultural geographies. It is beyond the scope of this text to do justice to these literatures although their influence is acknowledged in several sections of this book.

Within these economic, cultural and continental parameters, social geography is presented as a critical but changing social science, and as a purposeful and powerful opportunity to construct a field of valuable social knowledges. Attention is given to how these knowledges are socially constructed; how they draw on different **epistemological** approaches and practices; and how they are presented from different positions and for different purposes. This diversity is shown through the theoretical tensions discussed in Chapter 2 as well as the specific foci on social differences presented in Chapters 3–6. These latter chapters consider the core axes of social difference shaping contemporary social geographies: **class**, **gender**, **ethnicity** and **sexuality**. These differences support how we understand people's actions and multiple identities (across diverse categories of difference), and how places and spaces are involved in both shaping such differences and being marked by them.

But *Social Geographies: From Difference to Action* goes further than describing approaches to social difference. The third part of the book shows that beyond the specialist details that emerged in contemporary writings on difference there remain some thorny questions and opportunities. First, the potential and power of social geography can be seen when attention is turned from difference to the ways diverse people and groups will unite around particular issues or shared experiences. Despite differences, there are times and issues that appear to stimulate people to unite for temporary or ongoing reasons. Identity appears as one important concept for understanding this process. For example, people connect and assemble meanings and values around certain social categories, relations or experiences. An illustration of this process is given in Box 1.2. In this case, despite a variety of differences in parenting, sexual attitudes, ethnicity and economic circumstances, groups of women united under a lesbian identity to establish a range of alternative living spaces in the USA. Valentine's (1997) analysis of these activities shows how powerful a common (sexual) identity can be for achieving material and social alternatives for lesbians who are nevertheless a strikingly heterogeneous group of women.

Second, social geography can also advance by re-engaging with the opportunity to move beyond the specificity of difference, and recognize that geographies of power permeate the complexities of all social relations and difference. These issues are addressed in Chapters 7–9 which trace three ways in which social geographers move from the particularities of a

Box 1.2 Making lesbian space

Gill Valentine's work on geographies of sexuality has included an account of how some lesbians in the USA established rural female-only communities as an alternative society in the 1970s and 1980s. These communal spaces were to counter the perceived oppressive and **heteropatriarchal** conditions of the 'man-made' cities. Valentine notes that these actions were attempting to create a positive social space to counter lesbians' past experiences. However, she also records that social differences between the women meant that a united lesbian identity was always needing to be negotiated as differences between women surfaced. She writes.

> US lesbian feminists ... [produced] their own very different sort of 'rural idyll' – non-heteropatriarchal space – through the spatial strategy of separatism. By constructing the rural as an escape from the 'man-made' city these women draw upon stereotypical representations of the rural as a healthy, simple, peaceful, safe place to live. ... [These] attempts to create idyllic 'communities' by privileging the women's shared identities or sense of sameness as lesbians, over their differences. (1997: 109, 110)
> Separatists established land trusts to make land available to women for ever. This control of space, they believed, was essential because it would give women the freedom to articulate a lesbian feminist identity, to create new ways of living and to work out new ways of relating to the environment. (1997: 111)

A number of issues challenged the constructed unity lesbians tried to maintain. Valentine (1997: 114–17) documents these conflicts:

- Boy children: some lesbian lands banned all boy children and therefore excluded some lesbians because they were mothers of boys.
- Sexual and personal relations: different residents in lesbian lands experimented with celibacy, monogamy and non-monogamy but different personal choices affected women's ability to mobilize support for other decision-making processes in the communities.
- Class and financial resources: conflicts over ownership of land and ability to improve it (e.g. electricity and fences), and sources of income created cleavages between some women.
- Racism: while some lesbian lands drew on Black Power political ideology and strategies to succeed, the dominance of white women and claims of racism showed the fragility of these communities – and the limited access for African-American, Jewish and native-American women.
- Prejudice against disabled women: the lack of accessibility of some lands and the emphasis on physical labour reduced the way some women could participate in the lands.

In conclusion Valentine states:

> Lesbian separatists are one example of a marginalized group who have ... attempted to live out very politicized visions of a rural lifestyle, by emphasizing their shared identities. ... [But t]hese desires for mutual identification or homogeneity simultaneously appear to have generated boundaries and exclusions. (1997: 119, 120)

specific social variable (e.g. class or gender) to draw lines of relationship between various categories of social difference. By focusing on identity, power and action respectively, these chapters illustrate that contemporary social geographies can acknowledge – but move beyond – social diversity to address the possibility of (re)presenting people's choices and actions as diversity is negotiated through engagements with identity and power. Thus the closing chapters of the book concentrate on the records of recent social action that geographers have made, as well as synthesizing how social geography may continue to be a critically reflective means of seeing the uneven ways societies use and rework the environments and specific places in which they live.

1.2 WRITING FROM 'SOMEWHERE'

Before moving into the body of this book, it is important to emphasize that contemporary social geographies are particular accounts of what is often complicated and debatable content. They are projects that are placed at the intersection of diverse and sometimes contradictory concepts and experiences. At times these are expressed through dualisms that indicate the poles that might encompass the breadth of issues, for example, self and other, difference and unity, local and global. Nevertheless, an increasingly common quality within these geographies stems from the acknowledgement that all such knowledges are constructed from *somewhere*; that geographies are investigated and written by people who are working from specific personal and academic **positions**. This has been the case particularly for western/Anglo social geographies since the 1980s as knowledge and research choices have increasingly been recognized as culturally and politically situated (see Box 1.3).

It is appropriate therefore that some note is made of my position and interests in writing this book. I draw on specific academic identities to describe myself as a social, feminist and rural geographer. These labels – together with an interest in poststructural theories – locate most of my interests in geography. These include investigating how social theories may apply to the material and cultural worlds of rural Australia and New Zealand: how concepts of identity, difference and social movements play out in specific places (Liepins, 1998a; Panelli, 2002a; Panelli et al., 2002); and how socio-economic processes and cultural relations affect the social arrangements of rural households, farms, communities and industries (Liepins, 2000a; Liepins and Bradshaw, 1999; Panelli, 2001; Panelli and Gallagher, 2003). Throughout these considerations I have maintained an interest in critiquing some of the power relations and discursive processes by which social differences are established or maintained to the disadvantage of some groups (e.g. women, farm families, young people). Additionally, as an 'Australian' working in New Zealand and publishing in

Box 1.3 Examples of geographers writing from somewhere: reflections on position

Robin Kearns has maintained a long interest in the social and political issues surrounding health services. Working within a bicultural and multicultural situation in Aotearoa/New Zealand, Kearns has reflected on the way he has conducted geographic research into the provision of health services in Hokianga. He writes:

[T]he emplacement of Hokianga within my personal biography cannot be divorced from my place within the narrative of Hokianga's health system. I have visited the area since childhood, and have a number of personal friends in the district. Furthermore, at the time of beginning research, I had recently returned from a period in Canada where I had spent time in First Nations communities. My empathy to indigenous welfare issues, as well as awareness of my own cultural hybridity, thus contributed to (re)visioning a relationship with Hokianga people. (1997: 5)

Gill Valentine has experienced a widely accepted identity as a 'lesbian geographer'. However, in 1998 she wrote of the difficulties around maintaining multiple identities, especially in the face of harassment. In her account, she notes her different professional and personal positions as she reflects on the ways she was seen prior to her experience of victimization. She writes:

Although the choice of geographies of sexualities ... as a research topic at the beginning of my academic career was largely motivated by my own personal experiences ... I never set out to 'come out' within the discipline. ... [Nevertheless] the discipline [of geography] some-what thrust the identity 'lesbian geographer' upon me. ... [This] has placed me in a rather paradoxical position, for while I have been held up as a 'lesbian geographer' who is assumed to be 'out' both publicly and privately, I have actually been performing a very different identity to my family, creating a very precarious 'public'/'private', 'work'/'home' splintered existence. (1998: 306, 307)

Paul Routledge has investigated a range of geographies 'in action' where he has participated in political activities and written social, cultural and political geographies about these events. In one case, he writes of the social protests and geographies of resistance that sur-rounded the development of the M77 motorway in Glasgow, Scotland. He opens a paper on this issue by quoting from his personal journal, and then goes on to make his position(s) clear:

This journal entry refers to one of my personal experiences within the recent campaign against the M77 motorway extension in Glasgow, Scotland, which represented the country's first anti-motorway ecopolitical conflict. I participated in the 'No M77' campaign as a member of Glasgow Earth First! (one of the groups opposing the road) and Pollock Free State – an 'ecological encampment' that was constructed in the path of the projected motorway and which acted as the focal point of the resistance. ... In this paper, I want to examine the direct action component of this resistance. ... My analysis draws from my participation in the campaign. (1997b: 360)

Geraldine Pratt has conducted many years of research into the connections between work and gender. Recently, her work with Filipina domestic workers has sought to redress the fact that in her earlier work she had omitted the acknowledgement of such workers. She reflects:

[In the past] as a white middle-class academic I simply did not see the geographies of Filipina identity at one point in time. ... [Now] I see our job as one of creating trouble ... by making visible boundary constructions and the production of difference, and by keeping alive the question of who, inevitably, is being excluded as identities are defined. My current research involves an effort to make visible the boundary that prevented me from seeing domestic workers living in Vancouver in 1992. (1999b: 152, 164) international/western journals, I am conscious of both the influence and status of the theoretical 'metropolis' of the North (Berg and Kearns, 1998) and the opportunities for observing and constructing geographies from the margins (Monk and Liepins, 2000). Finally, having had the advantage of a postgraduate experience where a critical engagement with geography was supported within a wider departmental environment that encouraged and respected postgraduates' thought, dialogue and writing, I have sought to maintain the importance of listening across the spectrum of geographic voices. This is manifest in the book through my use of materials generated from a continuum of writers that stretch from the 'biggest' names of geographic and wider social theorists to the work of 'unknown' students who have inspired me as they have tackled social geography at the University of Otago. Student reflections from class sessions or reading logs are included in a variety of forms throughout the book. These diverse positions and interests of mine result in a wish to see social geography as a sub-discipline that can be open to constructions from many positions while recognizing the difficulties and hierarchies that exist in generating knowledge within the academic arena (Johnston, 2000). The book forms both a celebration and a critical reflection of social geography and I trust that it will encourage readers to always question:

- What are the currently dominant forms of knowing?

- How can we highlight inequalities, challenges and alternatives?

- What do we think are the important questions for social geographic inquiry?

- When and how do we respect and acknowledge other voices and knowledges?

- How can questions and problems be best investigated, understood and represented?

- What are the consequences and opportunities arising from our choices?

2 Contemporary Social Geographies: Negotiating Science, Theories and Positions

2.1 INTRODUCTION

The variety of contemporary social geographies being written engages with several of the questions that concluded Chapter 1, namely:

- What are the currently dominant forms of knowing?
- How can questions and problems best be investigated, understood and represented? and
- What are the consequences and opportunities arising from our choices?

Geographers regularly negotiate these questions, choosing to engage with different forms of knowledge and approaches to research. They position themselves and their work in particular theoretical and cultural ways which both affect the geographies being written and the consequences and opportunities that follow from such work (Barnes, 2001). The choices being made in such processes are vast.

Gaining a sense of the complexity of contemporary social geographies requires a critical appreciation of how these geographies are written, including the contexts and tensions that shape them. This chapter presents one perspective on these complexities. It argues that to understand the diversity of social geography we need to recognize the ways geographers have tackled decisions about science, theory and position. Such decisions are all highly political for 'what constitutes knowledge, that is, those ideas which gain currency through books and periodicals, is conditioned by power relations which determine the boundaries of "knowledge" and exclude dangerous or threatening ideas and authors' (Sibley, 1995: xvi). Sibley has documented a variety of excluded geographies as certain types of knowledge have held a dominant position in the discipline at different times in its history. To recognize how some views dominate, and others are marginalized, an understanding is needed of how ideas about **science**, **theory** and **positionality** shape the sub-discipline. The following four sections

11

provide an overview and illustration of these dimensions, indicating how different forms of knowledge have been valued at different times and for different reasons.

The following two sections focus on different views of science and theory. Then, because geographers rarely work through separate processes, deciding their views on science and selecting discrete theories, the fourth section shows that the process of constructing social geographies is often interwoven, personal and dynamic. This is not to suggest that geographers are *ad hoc* in their approaches to social inquiry. Rather, geographers and their work cannot be neatly boxed into (or kept within) certain categories of science or schools of theory. Social geographies are constructed through the interplay of different perspectives and debates, and the research biographies in this section illustrate some of the variety that tempers geographers' efforts to build meaning in systematic and purposeful ways. The variety of rural and urban research undertaken by Paul Cloke and Geraldine Pratt illustrate how dynamic (often challenged and reconstructed) these endeavours are. The fifth and final section raises the generic theme of positionality, showing how contemporary social geographies are becoming increasingly aware of the positions from which research is conducted and knowledge is constructed. As with all geography, social geography has ever been a political and socially constructed project. This chapter closes by emphasizing how this continues to be recognized as geographers seek to increase their reflection on why and how they write geography.

2.2 SCIENCE: SPECIFIC CONSTRUCTIONS OF KNOWLEDGE

Contemporary social geography, in Anglo-American contexts, emerges from the discipline's history of constructing geography as a scientific endeavour. By referring to Anglo-American geographies here, I am concentrating on the dominant traditions of the sub-discipline although other ethnographic and anarchistic social geographies were sometimes written (Blunt and Wills, 2000; Sibley, 1995), and the term social geography has been adopted in European traditions as a synthesis of diverse specialisms (Holt-Jensen, 1999). In the Anglo-American sense, social geography has been actively constructed as a social science since social meanings have been created based on specific choices and beliefs about how scientific knowledge can be established. These actions are not only academic or philosophical, they are also cultural and political since the purposes and cultural contexts in which we work affect how we develop geographic knowledge. (This is elaborated upon in sections 2.3 and 2.4).

Social geography, as a social science has been greatly influenced by the dominance of scientific thinking in postwar western academia. Following the Second World War and the growth in popularity of science and technology, geography was increasingly presented as a **spatial science**. Much of the popularity of geography as a spatial science stemmed from Schaefer's (1953) call for geographic research to become more scientific. Consequently, in the 1950s and 1960s, geography underwent a quantitative 'revolution' as scholars constructed general laws and predictions (Kitchin and Tate, 2000). While some of the fervour for spatial patterns and laws is absent from contemporary social geography, geographers continue to create contrasting works because of the choices they make about **ontology** (how the world exists) and **epistemology** (what constitutes knowledge). Different sets of knowledge are constructed by using contrasting theories (see section 2.3), however, and social geography has been fundamentally shaped by its relationship to notions of social science.

Understanding social geography as a social science presents at least two possibilities. First, social geography can be seen as a specific form of inquiry (often termed **hypothetic-deductive**). This form of social science involves commitments to certain types of data and research processes that seek to formulate laws – or at least models – that can be tested (replicated or challenged). While physical sciences have built laws about biophysical phenomena, social sciences have been criticized for pursuing law-building endeavours since data and conditions of inquiry cannot be held readily in controlled situations, nor can subjective and non-sensory processes be eliminated when investigating human behaviour. Therefore, social science has often spent more time on developing models rather than laws where such models illustrate the relationship between phenomena in conceptual forms.

Second, social geography as a social science can be seen as a range of knowledges that variously share generic interests. These include common understandings and interests in a construction of academic knowledge based on a consistently structured research process (e.g. asking carefully formed questions, studying data systematically and building explanations or other forms of meaning). In considering these alternatives, this section presents four dimensions of social science. First, it highlights how ontological issues of what can be known shape social science. Second, it indicates some of the ways social science is valued as a form of inquiry producing 'results' of particular merit. Third, by considering research methods, social science is presented as a continuum of practices stretching from a 'full pedigree' science – based on specific formally structured processes of 'scientific method' – through to a constellation of approaches that nevertheless share generic understandings about processes of inquiry. Finally, this overview indicates how contrasting notions of social geography as a social science stem from different philosophies of science. This commentary of social science is not presented as a comprehensive statement. Instead, it highlights some of the ways scientific understandings are valued and structured as common research practices. (For other broad reviews see Holt-Jensen, 1999; Johnston, 1991; Kitchen and Tate, 2000; Marshall, 1985; Waitt et al., 2000.)

2.2.1 Social science as a specific type of knowledge

First, social science can be seen as a form of science that focuses more on human and social subjects and questions than on biophysical objects and problems. Lewins (1992: 5) states that: 'social science is the attempt to explain social phenomena within the limits of available evidence ... [it] is concerned with finding explanations or theories of the social world'. Such a description shows how social science addresses *social content* ('phenomena'), but creates particular *types of knowledge* ('explanations and theories') that are predicated upon other understandings ('evidence': meanings that are given to data such as observations). The potency of such a form of knowledge lies in the assumptions that we can 'know' our world; that we can understand, explain or theorize it.

2.2.2 Social science as a valued form of inquiry and results

A second dimension of social science rests in the way it is highly valued as a form of knowledge creation. For instance, social science has been described as 'a fruitful mode of inquiry' (Gould, cited in Marshall, 1985: 113). The term 'inquiry' highlights how social science is not only a form of knowledge but also an investigative *activity*. It involves actions based on a desire to study, to inquire or ask questions, and to 'find out'. To this extent there is an explicit focus or purpose in a social scientific enterprise, most often signified by the statement of a research problem or question. There is also a developed culture of meanings and practices which propose the most valuable way to conduct such inquiry (see next sub-section).

Gould's description of social science as 'fruitful' also conveys a sense in which social science is worth doing; thus we return to the idea of valued knowledge. As with many other sciences (especially in a climate of competitive government or commercial funding: Johnston, R.J., 1996) the focus often falls on to the 'fruit' or output or product of social science. Consequently, reports, conference papers, books and articles are valued for the apparent qualities of that knowledge. Moreover, valuable outputs are prized for their *validity*, *objectivity* and *rationality*, qualities that are believed to stem from the way the inquiry has been conducted (see Table 2.1).

2.2.3 Practice: social science as a pedigree method or as a general form of inquiry

A third dimension of social science focuses on the *practices* that create knowledge. These include recognized ways of thinking and answering

TABLE 2.1   VALUED ATTRIBUTES OF A SCIENTIFIC APPROACH

Attribute	Description
Objectivity	• Idea that knowledge can be acquired in a neutral fashion. • Research is seen as objective if it claims to present a balanced account that evenly covers different perspectives and does not express bias in favour of any particular position or view. • Researcher is seen as an expert who objectively collects and analyses data without reference to personal judgements or preferences.
Rationality	• Idea that knowledge is of great value when it is rational, or based on reason (where reason is a means of establishing knowledge/arguments by way of assembling interconnected premises). • Research is seen as rational where theory and empirical work are based on internally coherent reasoning. • Researcher is seen as a logical thinker who can build knowledge via carefully assembled arguments and explanations based on formal reason.
Validity	• Idea that knowledge or research results are sound, legitimate and reliable. • Research that is seen as valid if results can be agreed upon (e.g. through repetition of analysis) or where conceptual reasoning is sound. • Researcher is seen as a reliable thinker who can design and analyse data in theoretically and practically sound and verifiable ways.

Note that recent postmodern and poststructural perspectives contend that there is no universal objectivity, rationality or validity. Instead, these attributes are all social constructions that will hold specific meanings and values for particular groups and that differ in time and place.

questions we might pose. The method is characterized as a logical set of processes by which we seek and analyse data. There are several contrasting processes that might be employed but together the scientific method forms a rigorous and recognized exercise of curiosity (in terms of clearly formulated questions) and a system for proceeding (via a series of related actions).

The practice of social geography as a social science varies considerably. In its most classic (pedigree) form scientific method is recognized as the 'logical structure of the process by which the search for trustworthy knowledge advances' (Marshall, 1985: 113). This structure presents a consistent set of actions, which enable the formation of explanations, and generalizations, which can be further studied and tested. Such methods differ from idiographic approaches that were popular before the twentieth century and focused on the unique geographic explanations that could be induced from the collection, classification and ordering of observations in a particular place or time. In a move to develop beyond much of the empiricism of postwar human geography, Harvey (1969) argued that scientific explanation in geography should follow formal hypothetic-deductive procedures (see Figure 2.1). Such practices enable a systematic cycle of modelling, hypothesizing, testing, and verifying explanation in a way that can be replicated.

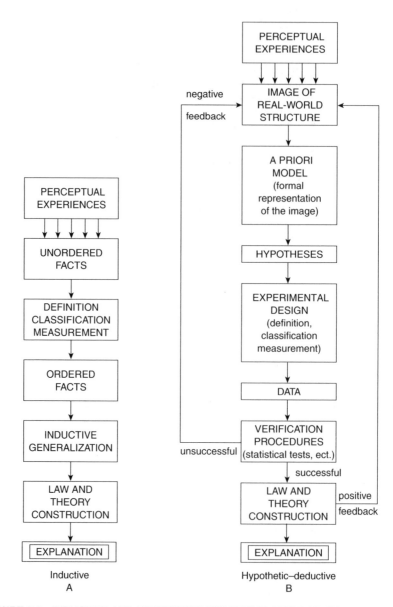

FIGURE 2.1 INDUCTIVE AND HYPOTHETIC-DEDUCTIVE APPROACHES TO SCIENTIFIC EXPLANATION

Source: Harvey, 1969: 34, Figures 4.2 and 4.3.

More broadly, social geography has employed the term 'social science' in a less purist form to communicate ideas about agreed general practices of inquiry. These practices include a belief in:

- professional conduct (with ethical standards and for an academic rather than personal purpose);
- identification of clear and relevant research questions;
- establishment of an appropriate approach to research; and
- implementation of a rigorous and repeatable system of collecting, processing and analysing data.

This broad view of social geography enables geographers to identify as social scientists not to practise a pedigree methodology but because of their commitment to, and valuing of, knowledge that is based on careful design, considerations of questions and structured forms of data collection and analysis. The variation in 'approaches' selected in this more general view of social science rests on the varying philosophies of science being adopted.

2.2.4 Contrasting sciences: philosophies of knowledge and inquiry

A fourth feature of social science that influences the diversity of social geography involves the contrasting philosophies adopted by geographers. Philosophical questions arise since valuing and forming scientific knowledge is not a singular or simple process. Instead, questions abound concerning both what can be known and how it can be known. This diversity rests on contrasting philosophies of science. A range of geographic texts has provided overviews of these philosophically different approaches to human (and social) geography. For instance, Jackson and Smith (1984) distinguish between positivism, humanism and structuralism; Johnston (1991) outlines positivist, behavioural, humanistic and radical geographies; and Unwin (1992), and more recently Kitchin and Tate (2000), employ Habermas's (1978) classification of sciences as empirical-analytical, historical-hermeneutic or critical.

While in each case the groupings and specialist forms within them vary somewhat, these classifications can be broadly distinguished in three ways (see Table 2.2). First, positivist and empirical geographies focus on the identification and measurement of objects that can be sensed. Knowledge is based on facts that are established by the orderly study of such objects. In the case of positivism, propositions are posited and laws sought concerning the relations between different phenomena. In this way such knowledges rest upon *realist* and *naturalist* ontologies.

Second, hermeneutic and humanist geographies focus on knowledge as interpretation rather than explanation and law-building. Knowledge

TABLE 2.2   THREE KEY PHILOSOPHICAL APPROACHES TO SOCIAL GEOGRAPHY

Philosophical approach to science and social geography	Realist or not?	Naturalist or not?	Questions of the social world	Subjects of research
	✓	✓		
Positivist and Empirical	A real world occurs irrespective of social constructions of it	The social world can be studied in a similar way to the natural world	What is it?	Material, sensate phenomena that can be observed or measured in some way
	✗	✗		
Hermeneutic and Humanistic	The world exists in the mind and 'reality' is constructed through thought and language	The social world cannot be studied with scientific method because it does not effectively recognize non-empirical evidence: ideas, values, etc.	What does it mean?	Actions, ideas, values and accounts that can be collected and interpreted
	✓	✗		
Radical and Critical	A real world occurs irrespective of social constructions of it	The social world cannot be studied with scientific method because it does not effectively recognize non-empirical evidence: ideas, values, etc.	What ought it to be?	Material and immaterial structures and relations that can be observed and critiqued, and possibly changed

production is less based on object-oriented facts and more involved with systems of meaning, communication and language. Consequently, such investigations can go beyond the material world to study people's experiences, values, actions and thoughts. In some forms (*existentialism* and *idealism*) these knowledges are both *anti-realist* and *anti-naturalist*.

Third, critical and radical geographies focus on why or how conditions exist and include both material and immaterial phenomena such as living conditions and power relations. Knowledge production is (implicitly/explicitly) associated with identifying structures and relations, which underlie and reproduce the social world. The radical dimension of such knowledge lies in the possibility that once understood, these structures and relations can be challenged or changed. Often these approaches (e.g. *historical materialism* and *realism*) are *realist* but *anti-naturalist*.

It is important also to identify the growth of postmodern and poststructural geographies. Debate surrounds their classification in different typologies of human geography. This includes problems raised by recognizing the deconstruction (sometimes rejection) of science that exist in postmodern and poststructural writings – yet typologies such as that suggested by Kitchin and Tate (2000) classify these schools of thought as 'critical science'. The radical and critical energies of these approaches have been recognized and consequently categorized as such by other authors (Holt-Jensen, 1999; Johnston, 1991) but, for the purposes of this text, I present postmodern and poststructural geographies as separate to radical geographies because of contrasting notions and approaches to the formation of knowledge. In short, while positivist, humanist and critical constructions of knowledge may all at various times be recognized as forms of social science, postmodernism and poststructuralism present major alternatives to scientific knowledge and highlight the importance of considering all knowledge as positioned or situated within specific conditions and terrains of power. They also often differ regarding the purposes of knowledge construction that may not be as focused on social or political change as the cases of Marxist and feminist thought.

2.3 THEORY: MAKING MEANING AND/OR EXPLANATIONS

As with other sub-disciplines of human geography, social geographies are written explicitly or implicitly with attention to theory. Theories take on different forms depending on the philosophies and epistemological beliefs being held. Johnston (in Johnston et al., 2000: 826) classifies these as positivist, idealist or realist. However, most geographers work from within more particular schools of thought as will be shown below.

Put simply, theories are series of linked statements, ideas and/or deductions that cumulatively form a coherent and distinguishable set of explanations or meanings of the world. However, such theories will vary with different philosophies (see Table 2.3). In some cases, theory is explicitly written to build logical 'explanations' that indicate (and sometimes model) how the social world operates (e.g. positivism, and behaviouralism). In other cases, theory is taken as a relative phenomenon. Explanation is questioned and critiqued as a limited endeavour and theory is presented instead as a means to construct specific and positioned 'understandings' or 'readings' of the social world (e.g. phenomenology, existentialism and postmodernism).

Theories become complex because of the difficulty in constructing understandings of the (multifaceted and ever-changing) social world. Consequently, each body of theory develops and employs language in quite complicated and particular ways. Detailed commentaries and examples of

TABLE 2.3 MAJOR FORMS OF THEORY ESTABLISHED UNDER DIFFERENT
PHILOSOPHIES OF SOCIAL SCIENCE

Philosophy of social science	Perspective and activities in developing theory	Outcome
Positivism and Empiricism	Theories can be developed as universal laws and may assist in predicting the future. Social scientists: • identify hypotheses and constraining conditions; • attempt to validate empirically; and • if validated, relations become laws that are assumed to be universal and are available for further testing.	
Hermenuetic and Humanist	Theories are individual and contextual constructions not universal. Social scientists: • identify systems of meaning, understanding, experience and communication; • investigate people's experiences, values, thoughts (e.g. lifeworlds); and • establish interpretations or readings of these experiences which explicitly or implicitly recognize knowledges (of both research subjects and researchers) as situated.	THEORY: A framework of concepts and relations or a set of related statements used to explain or interpret the world
Critical and Radical	Theories can frame/conceptualize the reality of the past and present and provide appreciation of the future and of how society might operate and change. Social scientists: • identify frameworks of concepts and relations to apprehend reality; • check for coherence of framework in providing understanding of underlying structures and mechanism (e.g. capitalism, patriarchy); and • test practical adequacy and potential for changing society.	

Source: Adapted from Johnston et al. (2000: 826) and Kitchin and Tate (2000: 6–26).

different theories in social geography have been widely published and some suggestions for reading are made at the end of this chapter. This section, however, sketches out some of the different theoretical approaches that have influenced social geography, paying particular attention to the *goals*, *contexts*, *tensions* and *connections* of these (often contrasting) forms of meaning and explanation. As an initial guide, these issues are summarized in Table 2.4.

Many commentaries reviewing these approaches dwell on the contrasting philosophical bases and conceptual details involved. The result is an impression of a neutral setting, where approaches are selected on philosophical grounds alone. However, geographers work within specific contexts and dynamic conditions in which their choices about building explanation, or understandings, has been deeply affected by the broader social contexts and internal academic environments in which they have worked. Capel (1981) notes that every academic community develops its own norms and value systems. These beliefs can powerfully shape geographers' activities and arguments. For instance, speaking of the popularity of quantitative geography in the 1960s, Morrill (1993: 352) recalls: 'I was an ardent advocate of the "new scientific geography"'. It should also be noted that the academic environment includes the financial and management circumstances existing at the time. For instance, R.J. Johnston (1996) notes the conservative and competitive conditions that have increased as a consequence of changes to the reviewing and funding of UK universities. These circumstances affect the culture surrounding geographers, which in turn shapes the purposes and goals of the research (column two in Table 2.4). Academic contexts include the power to legitimate certain forms of knowledge construction and to cast other forms as 'dangerous' or inappropriate (Sibley, 1995). For instance, positivist approaches were closely related to postwar modernist values and western societies' privileging of science and technology as a means to achieve progress through planning for urban growth and economic development:

> The years from 1945 to about 1965 were a period of scientific and technological dominance. ... The problems of production had been solved, it was claimed, in that enough goods and services could be provided to satisfy all. The problems of distribution were still being solved, for as yet there was inequality of provision at all spatial scales. But these could be handled, it was argued, and the prospect of a prosperous and healthy life for all was widely canvassed. Academic disciplines were contributing substantially to this problem solving, by their own scientific progress ... advances in the social sciences were aiding in the management of success. (Johnston, 1991: 27)

In this setting, objective, scientific knowledge was valued and questions of subjective meaning and human experience (such as humanists pursued) found little support until the 1970s and 1980s when a major swing against quantitative geographies began. At a similar time, developments in Marxist and feminist geographies drew on significant social unrest during the 1960s and 1970s when increased attention was given to uneven distribution of economic prosperity and the sexist, cultural and political practices that existed in many sectors of western society. These circumstances enabled social geographers to assertively critique and challenge previous positivist approaches.

TABLE 2.4   THEORETICAL APPROACHES EMPLOYED IN SOCIAL GEOGRAPHY

Approach	Overview and goals	Contexts
Positivism and Quantitative Geographies (Statistical and Mathematical)	• Observation and measurement of social phenomena to establish patterns and relations that can be tested for generic relevance. • Statistical approaches focused on description while mathematical approaches sought to build detailed models to explain and predict spatial phenomena.	• Post-Second World War popularity of science and technology as a means to achieve development or progress. • Widespread planning and policy interests encouraged quantitative studies that identified social patterns and processes, and modelled or predicted future trends.
Marxism	• Observation of political and economic relations and structures are made to show the uneven and contradictory social and economic power relations that shape and explain the organization of society. • Marxist theory as a critique of capitalism acts as a point from which dialectic reasoning studies the processes and contradictions in different cases of state, economic and social organization.	• Marxist geographies grew in strength throughout the 1970s and 1980s following the cumulative effect of the slowing of post-war economic growth and the increasing recognition of: unequal living standards; exploitative labour relations; the rise in multinational corporations and a new international division of labour.
Humanism	• Observation and ethnographic analysis of social phenomena to build understandings of how people experience social and spatial life. • Hermeneutic approaches sought understandings of generic human essence and focused on how people construct meanings. Humanism also sought to highlight human agency.	• Dissatisfaction with the macro-scale of positivism and Marxism led to a re-engagement with phenomenology, existentialism and pragmatism. Academic theory was seen as idealist and far removed from the subjective realities and sense people made of their everyday lives and actions.
Feminism	• Feminist approaches critiqued the unevenly gendered nature of past science, theory, research methods and the social systems in which we live. • Mixing quantitative and in-depth qualitative data enabled identification of the relations and processes by which women are generally disadvantaged.	• Second-wave feminism from the 1960s onwards provided widespread social and political exposure of women's issues and status, which supported the gender critique occurring within geography.
Postmodernism and Poststructuralism	• These approaches argue against universal truths such as metanarratives of modernism + science (even patriarchy) and the unproblematized theories that have been developed to support these ideas (including positivist, Marxist and early feminist theory). • Recognition of social diversity, and the importance of identity. Recognition of language as the means by which our world and power structures are constituted. • These approaches note the multiplicity and positionality of all knowledge and provide deconstructions, readings and representations of society (rather than observations or explanations).	• Sceptism and at times apathy, following the limited success of previous revolutionary hopes and energies for social change that had accompanied social movements and radical politics of the 1960s and 1970s. • Rising importance of literary and cultural theory provided ways in which to challenge the dominant and constructed nature of knowledge and beliefs in (e.g. in science, economics, medicine, etc.).

Tensions or debates	Developments and connections with other approaches
<ul><li>Internal debate surrounded the best way to construct geography as a 'spatial science'.</li><li>Tensions existed between primarily empirical research and that which was more mathematically theoretical. The latter was deemed essential if human geography was to address the complexity and dynamics of social life.</li><li>Critics noted that positivism focused on patterns and relations, but not why they existed.</li></ul>	<ul><li>Factorial ecology grew from the combination of factor analysis with earlier social ecology studies of the Chicago School (that were based on pragmatism and social Darwinism).</li><li>Quantitative analyses of living conditions and gender differences supported the early work of radical geographies such as welfare geography and feminist geography.</li></ul>
<ul><li>Many internal debates divided Marxist approaches, including theoretical debates seeking to explain the economic and spatial processes of capital accumulation, the cyclic nature of capitalism, the growing globalization of capitalism, the strategies for class politics, and the failure of socialist and communist politics in various national contexts.</li></ul>	<ul><li>Connections with feminist critiques enabled socialist feminism to draw on Marxist approaches to indicate how capitalism was socially reproduced through both class and gender relations.</li><li>Postmodern and poststructural approaches have stimulated Marxist geographies to investigate more fully the cultural processes that construct and perpetuate (but may also unravel) the socio-economic relations underpinning capitalism.</li></ul>
<ul><li>Internal debate surrounded the differences between phenomenology, existentialism and pragmatism.</li><li>Tensions existed between the personal and subjective scales of humanist geographical inquiry and the more voluminous macro-scale inquiries of positivist and Marxist geographies.</li><li>Critics from radical Marxist and feminist circles noted that humanism would maintain existing social relations when radical research sought to support social justice and change.</li></ul>	<ul><li>Pragmatism's anti-foundational perspective that questions the permanence or possibility of truth and knowledge complemented later practices in postmodern and poststructural geographies.</li><li>Attention to personal lifeworlds and socio-spatial experience at the level of individual people and bodies supported later developments in feminist and postmodern geographies.</li></ul>
<ul><li>Numerous internal differences shape feminist approaches, based on various explanations given for the socially uneven position of men and women in society: liberal, socialist, radical, psychoanalytic and poststructural feminisms all suggest different causes.</li><li>More recent debate followed the development of postcolonial critiques of feminism as a white, middle-class, modernist endeavour of the West that was unable to see the complexity and heterogeneity of gender relations and identities (e.g. in African and Asian societies).</li></ul>	<ul><li>Socialist feminism combined Marxist and feminist approaches.</li><li>Postmodern and poststructural feminism has moved from universal categories of men and women to look at the diversity of gender categories, relations and identities in shaping and perpetuating gendered meanings and inequalities.</li></ul>
<ul><li>A major and ongoing tension occurs concerning the politics of postmodernism and poststructuralism: do these approaches provide new critical opportunities or do they constitute a new apathy and relativism which can thwart the strategically collective voices of marginalized voices and politics, e.g. politics around ethnicity, gender and sexuality?</li><li>A second critique of the 'post' approaches includes the challenge that they have criticized past approaches as metanarratives, yet set their own perspective up as an elite narrative that makes equally broad claims concerning truth and the nature of society.</li></ul>	<ul><li>Intersections with feminism as noted above.</li><li>Intersections with Marxist critiques of capitalism, to demonstrate the diversity of capitalist and non-capitalist economic formations and relations that shape people's lives.</li></ul>

Considering each approach in a little more detail we can see that *positivist* and quantitative geographies developed highly valued descriptions and models of social structures, patterns and behaviours. Examples of this work include the compilation of large databases of social and economic indicators that were used to determine the structure of urban centres and rural areas (Cloke, 1977; Cloke and Edwards, 1986; Herbert, 1968). More mathematically imbued examples included the simulation and modelling of social change, such as Morrill's (1965 and 1968) study of residential segregation and diffusion. While these geographies were steeped in the scientific and philosophical values of positivism and empirical inquiry, it should also be noted they were not discretely or exclusively separate from earlier approaches. For instance Rees' (1971) review of factorial ecology drew on the heritage of social ecology research from the *Chicago School,* and indicated how these earlier ethnographic and pragmatic approaches could be expanded into large-scale studies of social structure and difference. For reasons of space, this book has focused on geographies since the 1950s. However, these have drawn on the heritage of past scholarship, including that of the Chicago School, which was particularly important for humanist geographers (for a detailed discussion, see Jackson and Smith, 1984).

Overall, positivist and quantitative geographies were full of technological and predictive potential yet increasingly they came under attack from other geographers for their limited explanatory power. Marxist geographers developed a critique reflecting the class-based social protests of the 1960s and 1970s, drawing on earlier labour movements of the nineteenth and early twentieth centuries. They pointed out that the 'spatial science' of the 'quantitative revolution' provided an increasingly elitist mathematical discipline that was incapable of explaining why existing patterns occurred and how they were maintained. In particular, patterns of uneven development and increased sensitivity to the failures of postwar 'progress' (such as social injustices around poverty, hunger and racial discrimination) resulted in geographers turning to Marx's theory of *historical materialism* to theorize a socio-economic critique of capitalist society. While Marxist geographies of the 1970s focused on capitalism, it should be remembered that Marx's theories built a critique of different *modes of production* and the class and labour relations within them. Geographers used Marxist theory to build a macro-scale explanation of social and economic conditions. They argued that the socio-spatial organization of capital created unequal conditions and development at global, national, regional and local scales because of the fundamental injustices surrounding production and labour processes and wider capitalist relations (Figure 2.2).

Marxist geographers worked across many sub-disciplines of human geography, producing both theoretical and empirical studies that were politically explicit in their critique of capitalism (Harvey, 1984, 1989). Theoretical work focused on the role of space, nature, class and the state in shaping and maintaining the socio-economic character of capitalism and

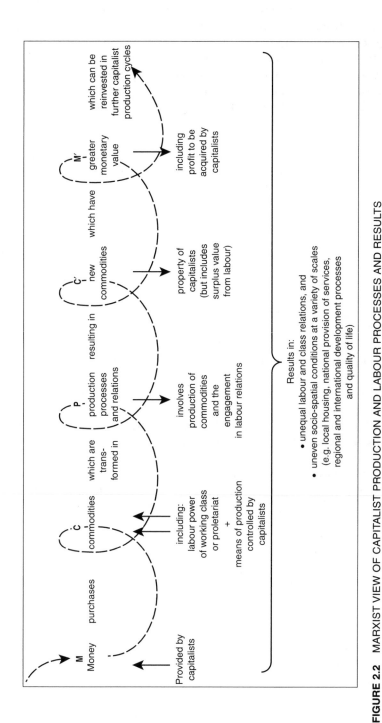

FIGURE 2.2 MARXIST VIEW OF CAPITALIST PRODUCTION AND LABOUR PROCESSES AND RESULTS

the class relations that supported it (e.g. Fincher, 1981; Harvey, 1976). For example, in discussing nature, Neil Smith (1984: xiv) argued that: 'In its constant drive to accumulate larger and larger quantities of social wealth under its control, capital transforms ... the problems of nature, of space and of uneven development are tied together by capital itself.'

Empirical studies included work across a number of scales. For instance, Dunford (1979) analysed the regional variations in capital accumulation across France, noting the social implications of these relations:

> Changes in the structure of the French space economy have been interpreted as ... the transition to a new phase of capitalist development. ... This transition ... has at the same time important social and political dimensions. It has been associated with the disruption of established local social and political structures, it has modified the class structure of the various regions and it has led to the formation of new class alliances. (Dunford, 1979: 103)

Alternatively, Massey (1984) developed a detailed argument concerning the importance of local contexts and social structures when assessing how production is spatially organized. Complementary studies of class structure indicated how changing social relations could be related to the dynamics of class formation and mobility (Cloke and Thrift, 1987). A further series of Marxist geographies identified the capitalist control of service provision. For instance, Smith's (1979, 1984) and Berry's (1981) analyses of US and Australian housing sectors argued that housing relations reflected ways in which capitalism is reproduced, and inequalities maintain important forms of capital accumulation. Across these observations and theoretical endeavours, Marxist geographies sought to explain the inequalities and injustices of social groups by recourse to a critical **metanarrative**, one which could both offer a political alternative and explain individual conditions via generic relations of production and social reproduction (Peet and Lyons, 1981).

In contrast, humanist geographies highlighted the diversity of human experience within the previously aggregated approaches of quantitative and Marxist geographies. These works have drawn on **hermeneutics** to note that human life involves more than positive/sensate phenomena or generic class and production relations, arguing that geographers should also be studying the 'world of everyday life' (Ley, 1978: 50). At a theoretical level, humanists turned to **phenomenology**, **symbolic interactionism** and **pragmatism** to support their work, while empirically they often undertook micro-scale qualitative studies of the experiences, ideas and values that shape people's understandings and choices (Ley and Samuels, 1978; Relph, 1970). In part, this provided an **agency**-focused counter-strategy to the structural Marxist explanations of social life according to broad class-based and economic-dominant critiques of capitalism. For instance, Seamon (1980a: 191) argued that geographers should study 'feelings and fantasies in relation to place; the role of the body in spatial behaviour; the importance of stability, continuity, and a sense of belonging in relation to one's environment'. His own work on people's everyday movement and 'place ballets' (Seamon, 1980b), and inner-city studies by Ley (1974) and Cybriwsky (1978), established a contrasting social geography that presented understandings of people's 'mundane', day-to-day worlds and actions. This approach constructed geographic knowledge not as a positively proven spatial law or prediction, nor as a grand scale historical-materialist critique of a mode of production. Instead it sought to construct knowledge as a series of understandings about essential human experiences and awareness of place that were derived from detailed observations, interviews and ethnographies. These techniques could accommodate data beyond the positive and measurable phenomena of earlier works.

Feminist geographies developed complementary geographies to both Marxist and humanist approaches. Feminist theory is enormously varied (for an introduction see Tong, 1992) and reflects the wider contexts of the second and third waves of feminism and the changing status of women in western societies. Feminist approaches in geography were based on the twin goals of critiquing gender inequalities within the discipline (e.g. McDowell, 1979; Momsen, 1980; Monk and Hanson, 1982) and developing a feminist research agenda (e.g. McDowell, 1992; Tivers, 1978). Feminist social geography drew on, but critiqued previous approaches (Rose, G., 1993). These geographies argued that all forms of social life (including economics, politics and culture) were influenced by unequal gender relations. While differences existed between socialist feminists and radical feminists concerning the causes of gender inequalities, all agreed that **patriarchy** constituted a specialized form of gender relations that disadvantaged women in western societies. Internal debates surrounded the emphases accorded to patriarchy and capitalism in understanding gender differences (see the debate between Foord and Gregson (1986) and McDowell (1986) and others in *Antipode*). Nevertheless, feminist approaches were to have a major influence on both the subject and practice of social geography. Feminist critiques of geography highlighted the need to review quantitative and qualitative research methods and reflect on the positions and power relations involving researchers and research subjects (Moss et al., 1993). Some alternatives included the further development of ethnographies employed by humanist geographies, as well as politically motivated critical methodologies aimed at more inclusive, participatory and action-oriented research where women subjects could more fully participate in research design and implementation for social change. Beside theoretical and methodological developments, feminist geographies also produced a range of detailed gender analyses of work, access to services, social relations and politics (see Chapter 4).

Postmodern and poststructural geographies have both critiqued positivist and Marxist approaches as universalistic and modernist projects. They also have challenged and enhanced feminist and humanist traditions in social geography. Akin to feminist and humanist geographies, they have also highlighted methodological issues for geography (e.g. Dear, 1994), paying special attention to the importance of language and **discourse** and the deconstruction of ideas and power relations within these frames of meaning. As with **pragmatist** and **feminist** views of science and knowledge, the 'post-' theories highlighted the socially constructed and time- and place-specific nature of knowledge. Postcolonialism is a further perspective that emphasizes these insights, for it critiques colonial discourses that have provided particular representations of the histories, people and relations that have been drawn together through western processes of colonization.

Drawing on wider contemporary cultural theory these geographies critique the search for universal explanations and the **metanarratives** of previous claims to truth (from conservative, positivist and quantitative approaches to social science through to the radical views of Marxism). They highlight the diversity of social life and knowledge, demonstrating the range of social experiences and discourses that shape place and space. Recognition of diversity has led to many social and cultural geographies that unsettle and deconstruct previously stable concepts such as 'city', 'rural', 'class', 'gender', 'race', 'scale', 'space', 'home' and 'community'. (See, for example, Fincher and Jacobs, 1998; Gibson and Watson, 1996; Gibson-Graham, 1996a; Halfacree, 1993; Liepins, 2000b; Rose, G., 1993; Smith, 1993; Soja, 1996.) As meanings and space have come to be seen as socially constructed and situated within particular contexts, issues of identity and representation (of meanings, research data and politics) have taken on much greater importance. They are important because they are seen to be negotiated, strategically assembled and performed both socially and spatially (Dwyer, 1998; Fincher and Jacobs, 1998; Forest, 1995; Pratt and Hanson, 1994).

2.4 NEGOTIATING THEORETICAL APPROACHES: DYNAMICS, INTERCONNECTIONS AND PALIMPSEST

Perhaps the most important observation one can make regarding theoretical approaches in social geography concerns the dynamic and interconnected character of these perspectives. The final column of Table 2.4 illustrates this fact, highlighting some of the ways different theories have been challenged or drawn into dialogue with contrasting schools of thought. In a similar fashion we need to realize that social geographers do not sit neatly or statically within different theoretical 'camps' or 'boxes'. Rather they, and the geographies they create, are constantly affected by the variety of theories in existence. They negotiate key differences and debates, and at times they alter their practice as a result of the tensions and new ideas that are stimulated within these schools of thought. Two examples illustrate this point.

2.4.1 Paul Cloke: from 'objective' indices through political economy to cultural readings and ethical reflections

Paul Cloke is a Professor of Geography at the University of Bristol and has constructed numerous social geographies associated with rural planning and change since the 1970s. His work illustrates the range of 'stories' (Cloke, 1994) that can be written as different theoretical approaches are considered and negotiated. Paul completed his training in the early 1970s when the dominant cultural values of his academic environment necessitated the acquisition and demonstration of technical and scientific skills associated with quantitative geographies. He explains that this context 'normalised modernity' and linked certain types of knowledge with the assumption that progress and planning were important research purposes:

> An emphasis in the Department ... was the idea that planning was both a legitimate target for social scientific endeavour and a process which was clearly capable of improving the situation of 'problem' areas. First in urban and regional contexts, then later in the rural environment, we were introduced persuasively to the notion of progress through planning to a better, more efficient world. ... [M]odernity became normalised in my geographical imagination at this early age, linking almost surreptitiously with the political interventionism of the time. (Cloke, 1994: 155)

Not surprisingly, this background supported Cloke's first statistical work: 'Responding to the cultural prompting that a social science thesis should display statistical competence, I spent a considerable amount of time ... to construct an "index of rurality" using principal component analysis' (1994: 156). This original index reflects many of the themes concerning measurement, objectivity and value raised earlier in the chapter (see section 2.2). For example, in introducing the index, Cloke stated:

> As it becomes necessary for special attention to be diverted to the planning problems encountered in rural areas, it is essential that attempts should be made not only to pin down the nebulous concept of rurality, but also to measure differences in the degree of rurality. ... The construction of an index of rurality for England and Wales is an attempt to formulate previously subjective ideas into one concise statement of rural–urban differentials. Statistical indexation techniques have been used in order to introduce as much objectivity as possible into the classification of different types of rural district. (1977: 31, 45)

The success of the index for academic as well as policy and planning purposes became somewhat embarrassing for Paul as the popularity and citation of the work was followed up with further examples of the index using new data (Cloke and Edwards, 1986; Harrington and O'Donoghue, 1998).

Tension developed for, like many other geographers, Paul was beginning to question many of the assumptions in his early work, and was reading and discussing more radical geographies (especially Marxist geographies focusing on political economy analyses that would enable him to reconceptualize his interest in rural change). By the late-1980s he was employing a critical approach to his research on rural change, especially the processes of capital accumulation, restructuring and class change. For instance, he argued:

> It is important to stress the centrality of **capital accumulation** as the driving force of social formation. The continuous process of reinvestment undertaken by individual capital units in search of surplus value is capable of generating unbalanced and unregulated trends in growth and decline. ... Restructuring does not take place in a societal vacuum. Different localities will have different histories of political and class conflict. ... As such, class relations are not only the end-product of foregoing rounds of capital accumulation and restructuring, but also serve to mould the characteristics of ensuring iterations of these processes. (Cloke, 1989: 181)

In this text, we see the foci and terminology of political economy geographies. These occupied Cloke's energies both in general commentaries of the socio-economic structure of rural areas and in the specific study and advance of detailed understandings – as in his argument for more nuanced intra-class analyses (Cloke and Thrift, 1987), and locally informed accounts of regulation (Cloke and Goodwin, 1992). The political economy perspective provided Cloke with a base for his socialist Christian approach to social geography (Cloke, 1994) and this illustrates the mix of theoretical and personal decisions that affect the constructions of geography.

Further developments in Cloke's work reflect the trajectories of social geography in the 1990s and 2000s. While retaining the 'wide-ranging notions of "truth" and "good"' in his approach to geography, Cloke reports that personal experience (of living in rural Wales) and contact with 'friend and erstwhile colleague Chris Philo' stimulated his interest in the more recent developments in social geography. These include the attention being paid to the social constructions of meaning and processes of marginalization that are found in postmodern and poststructural geographies. While Cloke clearly states that 'Christian and political impulses are too strong to accept the pluralist position' (1994: 187), his work demonstrates the hallmarks of these more recent geographies:

- readings of the social construction of meaning – namely rurality (Cloke, 1997);
- deconstructing notions employed in different discourses – especially focusing on poverty (e.g. Cloke, 1995); and
- providing attention to, and increasingly **reflexive** research of, 'Others' – namely homeless people (e.g. Cloke et al., 2000).

Examples of these latter works illustrate what Cloke (1994) has described as '(en)culturing political economy'. We see the overlapping influence of more than one approach to developing Cloke's social geographic knowledge. This has been the experience of numerous geographers and has at times been described as a palimpsestual process when different practices and meanings overlay each other but continue to shape the whole experience and ongoing conduct of geography (Cloke, 1997).

2.4.2 Geraldine Pratt: from social interactionism through Marxist and feminist analysis to poststructural accounts of difference, identity and research performance

Geraldine Pratt is a Professor of Geography at the University of British Columbia, Canada, and has constructed diverse social and economic geographies focusing on housing and work, and class and gender in urban settings since the 1980s. Her work illustrates the 'research performances' (Pratt, 2000b) that geographers negotiate over time and in different circumstances. Geraldine came to geography during her post-graduate studies after initial training and publication in environmental psychology (Russell and Pratt, 1980). In geography she was initially 'influenced by Jim Duncan's interests in social interactionism and the kinds of qualitative methods associated with Glazer and Strauss's grounded theory ... [and was] totally enthusiastic about being freed from the constraints of positivist psychology and experimental methodologies' (Interview, 2001).

As noted earlier in this chapter, both humanistic and Marxist approaches established strong challenges to positivist and quantitative geographies by the 1980s. The humanist tradition shaped the intellectual climate Geraldine first wrote from: 'The environment in the Geography Department at UBC was very exciting for me at the time that I fled psychology. The book *Humanistic Geography* [Ley and Samuels, 1978] had just been published and there was considerable interest in symbolic interactionism, phenomenology and existential philosophy. [And] a kind of anti-Marxism, anti-structuralism propelled debate around theories of class' (Interview, 2001). In this setting, Geraldine investigated class as a key axis of social difference. Subsequently, her first single-authored article (Pratt, 1982) reflected this context and 'engaged a theoretical debate among Marxists and Weberians about the status of domestic property classes and it allowed [her] to think through theories of class' (Interview, 2001).

Once Geraldine took up a position at Clark University she notes 'Marxism could come into perspective in a more reasoned and reasonable way ... [and] became more real and significant'. This context informed her continuing work and publications on class and housing (Pratt, 1986a, 1986b, 1989). While questions about class and housing and home spaces have remained an enduring interest in Geraldine's work, she notes that the other crucial context shaping the development of her work was her encounter with feminism and notions of gender:

[At Clark University t]he second influence was Susan Hanson. I really discovered 'gender' and feminism at Clark (and other women academics!). Gender and feminism were not unheard of at UBC but I wouldn't say it was much appreciated and was definitely entirely subordinated to class analysis ... but this focus [on gender] was firmed up on a day-to-day basis at Clark, particularly through Susan's direct mentoring. She drew me into a project on gender, labour markets and commuting. The project became more than that and increasingly it reflected both our shared and somewhat different interests. (Interview, 2001)

As will be detailed in Chapter 4, feminist geography grew rapidly in the 1980s as women geographers recognized gendered questions of social difference and political struggle in both their own lives and their research. Geraldine's collaboration with Susan Hanson, and her later work with migrant women, saw her make important contributions to feminist geography that linked gender with home and work (Hanson and Pratt, 1988, 1995; Pratt, 1999a; Pratt and Hanson, 1988, 1991).

But geography and Geraldine's experience of it were changing. Like Paul Cloke, Geraldine's reading of postmodern and poststructural literature stimulated her work and broadened her theoretical interests. Consideration of feminist and wider writings on poststructuralism and research methods resulted in Geraldine developing an increasingly reflective and eclectic approach to her work (Pratt, 1993, 2000b):

I have worked for the last six or seven years on domestic workers who come to Canada from the Philippines on a temporary work visa. This has been a collaborative research project with an activist group. I work with an eclectic range of theories and am most interested in work that pulls poststructuralist theories together with political economy perspectives and theories of radical democracy. [And] I am interested in experimenting with research methodology and writing strategy. (Interview 2001; see also Pratt, 1997, 1999b)

Her interests in home and work and class and gender continue, but they are now much more consciously and critically written in recognition of the way social difference intersects with identity formation; the ways identities and choices are discursively and materially shaped to advantage some and not other groups; and the way geographers can write geographies that open opportunities while continuing to document the boundaries and differences that separate us. She notes:

... [O]ur commitment should be one of opening doors for communication and ... this can be done not only by documenting the hybridity of cultures (that is, the fragility of borders and the interdependence of differences) and the creative potential for new critical identities to emerge at the border ... but remembering the exclusions that found every identity. I see our job as one of creating trouble ... by making visible boundary constructions and the production of difference, and by keeping alive the question of who, inevitably, is being excluded as identities are defined. (Pratt, 1999b: 164)

These sketches of Paul Cloke's and Geraldine Pratt's work as geographers are not comprehensive biographies or commentaries on their publications. Instead, they are presented to show the dynamic nature of geographers' research and writing. While entering geography during different 'eras', both individuals have experienced the broad theoretical currents moving through geography. And stimulated by their colleagues and students, they have found connections between theoretical approaches and built-up palimpsestual understandings of social geography – where, just as in geological rock strata, theoretical approaches overlay one another within a sub-discipline (and an individual geographer's work) and the contemporary 'surface' of geographic practice is nevertheless influenced by the previous strata/approaches laid down in earlier times.

Every history or biography is selective and my choices here have been premeditated. For me, Cloke and Pratt show how geographers with diverse backgrounds can pursue valuable studies of social difference, can question inequalities and injustices, and can reflect on their own work and the possibilities that geography offers in terms of other ways to imagine, understand and write the social world. For me, these examples make social geography worth pursuing. They also highlight the different **positions** the two geographers have (in geographic heritage, gender, sub-disciplinary specialism and methodologies) which links to the final consideration of social geography as a negotiation of positions.

2.5 SOCIAL GEOGRAPHY AS A NEGOTIATION OF POSITIONS

This chapter has shown how social geography varies because of the different ways geographers have understood and practised the subject as a scientific and theoretical endeavour. Differences in approach have also highlighted the variety of contexts and motivations with which geographers work. These situations affect the **positionality** of the geographers, including the choices they make in topics to investigate, ways to conduct research, and means they choose to interpret and present their results. Explicit consideration of positionality has increased as recent perspectives such as feminism, postmodernism, poststructuralism and postcolonialism have emphasized the subjectivity and politics involved in constructing all forms of knowledge. For instance, humanists argued that geographers should 'examine their own credibility' (Ley, 1978: 49), while feminist geography more radically illustrated the partial (especially gender biased) ways in which geography was being constructed (Rose, G., 1993).

More recently, postmodernism, poststructuralism and postcolonialism have noted the diversity of knowledges that may be constructed using different discourses and perspectives. Traditional scientific knowledge is recognized for the privileged yet particular worldviews it projects, including both the findings made and the status of researchers and research subjects. These 'post-' critiques have resulted in wider considerations of the highly constructed qualities of knowledge and the personal and wider politics associated with research design and representation. Penrose and Jackson (1994: 208) note: 'Our status as scholars places us awkwardly with respect to the hegemonic system which we may wish to resist. ... All scholars are faced with the choice (whether they recognize it or not), of using their power either to reproduce or to challenge hegemony.' The cases of Paul Cloke and Geraldine Pratt and their recent work with rural homeless and urban Filipina domestic workers shows ways in which hegemony can be questioned. Claims of objectivity are being laid aside as social geography is increasingly imagined as a self-consciously positioned and reflexive social science where scholars take explicit 'speaking positions' (Pratt, 1992). Crang's reflections on cultural geography are equally pertinent here for social geography. He notes:

> It is not the case that social factors (our backgrounds, the context of research) can either be factored out, or that they devalue our knowledge; instead such tacit or practical factors are vital in creating knowledge. They cannot then be simply removed as though they contaminate or corrupt the work. Scientific knowledge should not be seen as being contaminated or 'biased' by social factors; rather science should be seen as a social process. ... The situatedness of knowledge in and between the cultures of researcher and researched highlights the importance of thinking about why we carry certain assumptions and connecting our biographies to what we study. (Crang, 1998: 184–5)

The cumulative effect of these considerations has meant that social geographies can be considered increasingly critical and reflexive. England explains that this involves:

> ... self-critical sympathetic introspection and the self-conscious *analytical* scrutiny of the self as researcher. Indeed reflexivity is critical to the conduct of fieldwork: it induces self-discovery and can lead to insights and new hypotheses about the research questions. A more reflexive and flexible approach to fieldwork allows the researcher to be more open to any challenges to their theoretical position that fieldwork almost inevitably raises. ... And the reflexive 'I' of the researcher dismisses the observational distance of neopositivism and subverts the idea of the observer as an impersonal machine. (England, 1994: 81)

The following chapters focus on social difference, identity, power and action, and indicate the contrasting constructions that have been made as a result of different academic positions and interests. This includes the possibility of challenging past 'geographies of exclusion' and creating more heterogeneous, inclusive and 'egalitarian' accounts of society (Sibley, 1995). In contemporary social geographies the expectation is building that geographers will more critically reflect on the way academic knowledge is constructed and develop practice that more fully appreciates different positions, experiences and meanings. Geraldine Pratt's recent writing, cited in section 2.4, illustrates this type of move and Sibley's argument for an explicitly positioned practice is an encouraging statement of these possibilities:

> Understanding the experience of others and their relationship to place involves positioning ourselves in the world. Listening to and talking with people is one necessary part of this endeavour. Reflecting on the experience in such a way that we recognize our own part in the dialogue is another. (Sibley, 1995: 186)

SUMMARY

- This chapter has outlined the diverse character of social geography as a field of contrasting approaches and socially constructed knowledges.
- Differences rest in the choices and tensions that shape geographers' understanding of and engagement with 'science', theories and a diversity of academic and social positions.
- The dominant form of knowledge construction in social geography in the latter half of the twentieth century involved social science.
- Social science developed from broader views of science as a particularly formal way of assembling knowledge based on the development and testing of explanations.
- Scientific method was valued for its perceived objectivity, rationality and validity.
- Varying approaches to social science developed as a result of different philosophical perspectives. In each case, the politics of valuing certain ontologies and epistemologies is linked to the practices of how social geographic research is investigated and presented.
- Social geographic theory is varied but can be understood as a way to establish different explanations or interpretations or critiques or readings of the social world.
- Quantitative and positivist approaches saw the rise of geographic explanation, humanist approaches resulted in geographic interpretations, Marxist and feminist approaches produced geographic critiques, and postmodern/poststructural approaches have seen the increasing popularity of geographic readings and deconstructions.
- The diversity of theory stems from the different philosophical perspectives geographers adopt. However, these cannot be seen as discrete and mutually exclusive positions. Rather, social geographies

reflect the interconnections, palimpsest and cumulative influence of different approaches. Geographers are affected by existing research and knowledge and engage with, challenge or extend different perspectives so that interconnected layers or strata of knowledge and geographic practice build up over time for any given geographer or topic (e.g. poverty, class, etc.).

- Contemporary social geographies reflect the theoretical stimulation that has resulted from debates and engagements within and across different schools of thought.
- Contemporary social geography continues to value structured forms of inquiry, but also now combines an increased recognition of the need to reflect on the position and politics shaping the purpose and practice of social geography.
- Acknowledgement that geographers always write from particular contexts and positions has encouraged a more reflexive practice where many geographers now closely consider how their background, practice and writing of geography are intertwined.

Suggested reading

Useful introductions to different theoretical approaches in geography appear in a variety of forms. A long, detailed account is provided in book form by Cloke et al. (1992); a short overview is provided by Kitchin and Tate (2000) in their opening chapter 'Thinking about research'; and Waitt et al. (2000) devote Chapters 2 and 3 of their *Introducing Human Geography* to a review and explanation of the common and recent forms of geographical explanation. Specific comparisons between Marxist and humanist approaches appear in a paper by Ley (1978), and an important account of exclusion and reflexivity in geographic knowledge is given by Sibley (1995) in the last four chapters of his book, *Geographies of Exclusion*. Finally, a key book-long, feminist critique of geography is given by G. Rose (1993). In terms of biographies and context that shape geography, a general text is provided by Johnston (1991) while specific individual biographies or autobiography and/or reflection on practice as geographers is given in papers and chapters by Barnes (2001), Cloke (1994) and Pratt (2000b).

CATEGORIES OF SOCIAL DIFFERENCE

PART II

In seeking to investigate and form understandings of social life and the environments in which it occurs, social geographies have commonly highlighted the notion of social difference. At different times, and in different ways, social difference has been conceptualized and investigated in terms of categories, relations and spatial patterns or struggles. Nevertheless, the most frequently studied categories of social difference have been **class**, **gender**, **ethnicity** and **sexuality**. This part of the book takes these four foci as a structure for exploring the social and spatial patterns and interlinkages that occur at a number of scales. Chapters 3–6 concentrate on an individual category but reflect many of the overlapping theoretical perspectives and approaches introduced in Chapter 2. Each chapter documents how categories of class, gender, ethnicity and sexuality have been conceptualized and investigated in the last 30–40 years. They highlight both common themes and the ways geographies have overlapped and encouraged greater attention to the coincidence and intersection of differences (both these four core differences and further ones as well, e.g. age, dis-ability, etc.).

Recurring throughout this work is an acknowledgement that social differences are not simply artificial categories. Instead they reflect real material, cultural and political divisions and inequalities. As a consequence, many geographers have chosen to focus on the social processes and relations that separate people (and mark places and spaces) on the basis of differences. Implicated in this work is the need to understand how inequalities are created and maintained. This has been tackled with attention to power but in contrasting ways, e.g. Marxist critiques of capitalist labour relations, feminist accounts of **patriarchal** gender relations, and **queer** theories on the **heteronormative** constructions and relations that dominate societies.

The following chapters detail the relations that unequally differentiate individuals, groups and societies along these four commonly recognized axes. They also record important attention that is being given to further differences (including additional categories of difference) and the way all expressions and struggles over difference are embodied, performed and embedded in spatial contexts and consequences. Thus, the end of each chapter raises a number of overlapping or further developments that are invigorating social geographies of class, gender, ethnicity and sexuality. These conclusions also respectively foreshadow the wider discussions of identity, power and social action that are taken up in Part III of this book.

3 Class

3.1 INTRODUCTION

How would you describe your class position in society at present? The questions displayed in Box 3.1 might be asked of you if a researcher was compiling a class profile of you or your household. Depending on the philosophical and theoretical approach guiding the research, this study may concentrate on your class position in terms of your economic situation and/or your social status and behaviour. Irrespective of approach, however, *class* has been one of the earliest categories of difference that social geographers have employed when describing and explaining unequal social conditions and experiences.

Both economic and social geographers have used concepts of class to help them identify, analyse or read the way different groups of people experience many of the material conditions of life. Class can be thought of in diverse ways but its dominant use is associated with aspects of life related to people's economic position. Put simply, people's access to income and exchange processes will affect the material (and cultural) shape of their lives. It will affect: where they live; the condition of their shelter or housing; how they spend their day (and night); whether they have leisure time (and how this is spent); and how they might relate to other people in the similar or different class positions.

This chapter looks in detail at the way class has been understood conceptually (section 3.2) and how geographers have investigated class over the past three decades (section 3.3). These sections show that ideas about class have ranged from classifications based on various economic and social characteristics, through to critical theories of class that have highlighted the unequal relations and conflict existing between classes. Recently, both definitions and geographic inquiries have been challenged by poststructural approaches where class is seen less as an essential social axis or discrete category and more as an *overdetermined* social process.

Box 3.1 A short survey

1. What is your annual income?

☐ < $20,000   ☐ $20,000 – 39,999
☐ $40,000 – 59,999   ☐ $60,000 – 79,000
☐ $80,000 – 99,999   ☐ $100,000 – 199,999
☐ > $200,000

2. How do you derive your annual income? (Tick all relevant boxes)

☐ Paid employment   ☐ Shares and dividends
☐ Superannuation   ☐ Unemployment benefit
☐ Single-parent benefit   ☐ Other (please list all sources)

3.   If you are in paid employment:

a.   What is your occupation?

☐ Legislators, Administrators and Managers   ☐ Professionals
☐ Technicians and Associate Professionals   ☐ Clerks
☐ Service and Sales Workers   ☐ Agriculture/Fishery Workers
☐ Trades Workers   ☐ Plant/Machine Operators and
☐ Labourers and Elementary Service Workers   Assemblers

b.   What is your employment status?
☐ Full-time employee   ☐ Part-time employee
☐ Casual employee   ☐ Self-employed

c.   Do you belong to an employment-related union?   YES/NO

4.   What is your highest level of education?

5.   Do you own any property or investments (e.g. land, rental property, shares)?
                                                              YES/NO

6.   Where do you live?

☐ Home owned by yourself and/or kin   ☐ Rental accommodation
☐ Other (please describe)

7.   Which of the following do you own?

☐ Car   ☐ Television   ☐ Sound system   ☐ Computer

8.   How do you spend your leisure time?

☐ Watch sport   ☐ Go to the pub   ☐ Shopping
☐ Participate in sport   ☐ Go to concerts/cinema/opera   ☐ Other, please state

---

This means that class is seen as a process that is mutually linked or constituted with other equally influential aspects and contexts (e.g. environmental, political and cultural aspects of life). These newer trends are identified at the end of section 3.3 and form the basis of the closing section of this chapter where contemporary challenges and connections surrounding class are tabled. A biography in Part V has been written to accompany this chapter, and using a similar format as that shown for Paul Cloke and Geraldine Pratt (section 2.4), the section presents a sketch of the multistranded approach that Andrew Herod has taken when studying geographies of class.

3.2 DEFINING CLASS: SOCIAL GROUPS BASED ON MATERIAL CONDITIONS AND RELATIONS

Class has developed as a classic category of social difference in human geography and sociology. The concept has been used extensively through the latter half of the twentieth century as scholars have sought to define and investigate how contemporary western societies are structured (Giddens, 1973). While other cultures and historical eras have been structured around different social stratification (e.g. castes and tribes), class-based considerations have led geographers to focus upon social characteristics and economic relations as they occur in modern capitalist and 'western' societies. In particular, class has been used to signify the distinctions between people on the basis of their possessions (especially property, capital and labour power) and their activities (especially various forms of work and lifestyle choices and political actions or class struggles).

The range of theoretical approaches introduced in Chapter 2 is also evident in geographies of class. Some descriptive, quantitative studies have been developed using measurable social-economic indicators of class. These gradational classifications have tended to be descriptive rather than explanatory (Johnston et al., 2000). However, most social geographies have concentrated on the relations that occur between different classes as a result of the economic structuring of society. As shown in Figure 3.1 'western' geographies, focused on class relations associated with capitalism, have highlighted class indicators (such as income, home ownership or education and consumption levels). They have also explained ideas of class in terms of the system by which societies are economically organized i.e. through market interests or **modes of production**. These two emphases appear respectively in the Weberian and Marxist approaches adopted by scholars writing relational geographies of class.

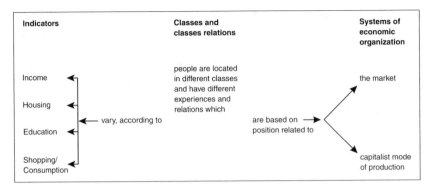

FIGURE 3.1   FEATURES OF CLASS ANALYSIS IN CAPITALIST SOCIETIES

3.2.1 Classes, life-chances and markets: Weberian approaches

Max Weber conceptualized class in capitalist societies in relation to people's market position, their interests and opportunities, or 'life-chances'. He wrote:

> The capitalistic economy of the present day is an immense cosmos into which the individual is born, and which presents itself to him, at least as an individual, as an unalterable order of things in which he must live. It forces the individual, *in so far as he is involved in the system of market relationships*, to conform to capitalistic rules of action. The manufacturer who in the long run acts counter to these norms, will just as inevitably be eliminated from the economic scene as the worker who cannot or will not adapt himself to them will be thrown into the streets without a job. (Weber, 1976/1930: 54–5, my emphasis)

Although this view was originally expressed in 1930, Weber's way of situating different classes of people (e.g. manufacturers, workers) within a market context was to prove an important basis from which to explain people's class position and their life-chances or opportunities and living conditions. An economic foundation for class noted how different groups of people brought their resources (including their capital, skills and ability to labour) to the market. These groups would then receive material and symbolic returns in terms of income and recognition. Consequently, Weber identified class in relation to economic life-chances. He explained class as:

> [T]he typical chance for a supply of goods, external living conditions, and personal life-experiences, in so far as this chance is determined by the

amount and kind of power, or lack of such, to dispose of goods or skills for the sake of income in a given economic order. (Weber, 1961: 181)

Weber did not see people's life-chances as random. They were understood as an outcome of people's position in relation to markets. Therefore, a focus on market conditions and processes was an important part of Weber's ability to differentiate people's experiences over time and space. The market provided individuals with the 'chance' to exchange their goods, skills or labour for income and/or other possessions that would sustain their lives.

Weber recognized different chances and conditions. Consequently, some of these classes were: the 'positively privileged classes' who were the propertied or commercial classes with substantial goods or assets to take to the market; the 'negatively privileged classes' such as paupers and workers who lacked property and/or skills; and the middle classes such as officials, craft-workers and professionals whose privilege or life-chances varied (Edgell, 1993; Saunders, 1990).

While Weber noted that economic class was a fundamental factor affecting individuals' material and social life, he also argued that class was only one of several key social differences. In ways that would later resonate with more contemporary attention to diversity, Weber contended that individuals' power (party affiliations and relations) and social status would also distinguish their experience of life and their ability to shape their society. Seeing these as separate characteristics and factors is one way in which Weberian class approaches differ markedly from the second, Marxist approach.

3.2.2 Uneven class relations in capitalist modes of production: Marxist approaches

Marxist approaches to class focus on the way in which societies are differentiated according to their position within a **mode of production**. In capitalism, therefore, class is established in terms of people's relation to capital and a capitalist mode of production. Marx argued that three broad classes could be distinguished:

The owners merely of labour power, owners of capital, and land-owners, whose respective sources of income are wages, profit and ground-rent. ... [I]n other words, wage-labourers, capitalists and landowners, constitute the three big classes of modern society based upon the capitalist mode of production. (Marx, 1967: 885)

In practice, Marxist analyses of class have focused on two broad categories where landowners and capitalists are conflated into one category (see

Table 3.1). On the one hand, the bourgeoisie own the means of production (property and capital) while the proletariat own only their labour.

TABLE 3.1   KEY CHARACTERISTICS OF TWO CORE CLASSES

Characteristics	Class 1 – frequently called: Capitalists Bourgeoisie Employers	Class 2 – frequently called: Labourers Proletariat Workers
Own:	Means of production: property, resources and/or capital	Labour
Derive income from:	Profits	Wages

While Marxist approaches to class, like Weberian ones, recognize differences in levels of wealth and property, class is considered primarily as a relational notion based on inequality and conflict. Attention is given to the uneven economic and social relations associated with the mode of production. In capitalism, classes are considered in terms of how groups are positioned within the mode of production, and how they relate to each other. Massey explains:

> The basic building blocks of a capitalist society, the bourgeoisie and the working class, form the main axis. Each of these two classes is defined in relation to the other and by its degree of control over the process of production, its place in the relations of production. The bourgeoisie is defined by having both economic ownership and possession of the means of production – in other words, 'capitalists' control the accumulation process through decisions about investment, decide how the physical means of production are to be used and control the authority structure within the labour process. The working class is characterized by the absence of these kinds of control, by its exclusion from both economic ownership and possession. These are the two fundamental, and defining, classes of a capitalist society. It is their mutually defining relationship which enables the one class to exploit the other and to do so in a specifically capitalist fashion. (Massey, 1984: 31–2)

Class relations and conflict are seen as central to Marxist analyses and geographers using this perspective have shown that these occur in different ways over space and with place-specific contexts. Different arrangements of production, divisions of labour and influence of state/political structures are seen to vary over regions and sub-regional spaces:

> [Capitalist class] relations occur over space. The changing spatial organisation of the relations of production and the division of labour

is a basis for understanding changing spatial patterns of employment and the geography of social class. (Massey, 1984: 39)

The detailed expression of these generic spatial patterns have been studied in a variety of locations and examples are given in section 3.3.1.

In sum, Weberian notions of class emphasize individuals' positions in relation to the market and the consequent life-chances individuals have regarding access to housing and employment. Geographers following this tradition can draw attention to housing and labour markets and the spatial patterns that develop and change over time. In contrast, Marxist notions of class emphasize uneven economic relations and conflict as a part of capitalist mode of production. Social geographers following this tradition focus on the spatially uneven capitalist relations in operation, noting the contexts, processes of class struggle and uneven living and/or working conditions.

3.3 CONTRASTING GEOGRAPHIES OF CLASS

Geographies of class have drawn on a number of traditions introduced in Chapter 2. Marxist, feminist and poststructural approaches have been employed most commonly by geographers studying class over the past three decades. This section outlines the range of studies that have developed since radical and critical perspectives gained popularity in the discipline through the 1980s. First, works identifying class groups and relations are reviewed, where geographers highlighted both the social and spatial processes involved (section 3.3.1). Second, studies of class change are discussed in both urban and rural cases, with note being made of the different class tensions and politics involved (section 3.3.2). Finally, poststructural geographies of class are introduced as contemporary authors have explored the discursive as well as material processes constituting class identities and shaping the ongoing study of class politics (section 3.3.3).

3.3.1 Identifying classes: groups, relations and spatial processes

Numerous studies in the 1980s documented class structure and class relations associated with capitalist economies. Some geographers and sociologists focused on documenting social class, related to social and residential areas within urban studies (Robson, 1975; Saunders, 1984). Mapping and description of different (low, medium and high) status areas were regular activities in this type of study. For instance, Robson's (1975) account of urban social areas enabled the classification of residential areas which highlighted status based on combinations of residents' occupation, income, education and lifestyle. His account of social segregation was shown to have a distinctly spatial form and these types of factor continue to differentiate residential neighbourhoods in many settings (see Figure 3.2).

FIGURE 3.2 CONSTRASTING SOCIAL CLASSES: RESIDENTIAL COMPARISONS

Source: Panelli, private collection.

In contrast, Marxist approaches to class often concentrated on labour relations associated with people's participation in the formal economy – for example, see the labour geographies that Herod has developed (in Part V). Participation in the formal economy involves individuals' positions in the capitalist production of commodities (goods and services) or the management of various circuits of capital. This distinction is made since feminist scholars quickly noted that all manner of other work and social reproduction continued in western societies beyond the formally recognized economy, but was rarely acknowledged economically, or by Marxist class analysts. Using a Marxist approach, considerable work was done in analysing the employment status of different populations to determine different people's positions regarding the ownership of capital or the participation in various labour markets. This type of work was advanced by geographers adopting Wright's (1979, 1985) extension of Marx's class categories. Wright's typology recognized a range of capitalist, middle and working classes, distinguished as class fractions based on different groups' positions in relation to employment, management of labour, and skill levels (see Figure 3.3). These distinctions enabled geographers to make more detailed studies of class differences across space, and to highlight class conflict both between and with class categories (Cloke and Thrift, 1987; Massey, 1984). The strength of Wright's creation of class fractions (or sub-categories) within the broad Marxist divisions meant that geographers like Cloke and Thrift (1987) could study the specific class relations and 'battle lines' of individual fractions, for they argued:

> Each of the locations in [Wright's] class structure can be thought of as a particular ensemble of class relations which people in particular locations can struggle to establish either uniquely or in alliance with people in other locations, distinctive classes and class fractions. ... Wright provides a map of where some of the main battle lines in modern capitalist societies should be drawn. (Cloke and Thrift, 1987: 326)

The majority of these geographies concentrated on the spatial unevenness and inequalities resulting from capitalist processes. For instance, Doreen Massey and Richard Walker published major pieces on class and the division of labour, noting them as both social and economic features of western societies. Massey documented the spatial division of labour in the United Kingdom, noting how regional histories and cultures would affect both class character and the relations occurring between labour and capital, for example: 'Compare the individualistic stroppiness of Merseyside workers, with a long history behind them of casual labour on the docks, with the organized discipline of miners from South Wales' (1984: 58). In a complementary fashion, Walker recorded the temporal and spatial character of classes through changing phases of capitalism and employment arrangements. He linked class categories to wider dynamics of power that

	Owners of means of production	Non-owners [wage labourers]		
Owns sufficient capital to hire workers and not work	1 Bourgeoisie	4 Expert Managers	7 Semi-credentialled Managers	10 Uncredentialled Managers
Owns sufficient capital to hire workers but must work	2 Small Employers	5 Expert Supervisors	8 Semi-credentialled Supervisors	11 Uncredentialled Supervisors
Owns sufficient capital to work for self but not to hire workers	3 Petty Bourgeoisie	6 Experts non-managers	9 Semi-credentialled Workers	12 Proletarians

FIGURE 3.3   WRIGHT'S TYPOLOGY OF CLASS FRACTIONS
Source: Wright (1985: 88).

are played out in diverse spatial patterns based on differing labour processes, industries and histories:

> Class is a relation of power which must continually be maintained, extended and recreated in the face of changing conditions. It is therefore inextricably entwined with the development of capital(ism) as a system of production, circulation and exploitation. ... The location process thus becomes a strategic part of the employment of labour by capital. ... The result of the location process is, at any time, a mosaic of workplaces and associated communities. This mosaic is literally a spatial division of labour. That spatial division of labour is necessarily uneven in its development because of the differences among labour processes and the idiosyncratic element in workplace employment relations. And the mosaic shifts continually over time as industries evolve and labour relations are reconstituted, with the new overlaying the old in a rich criss-cross of industrial and class history. (Walker, 1985: 172, 185)

Following Massey and Walker, many geographers in the 1980s used the increasingly popular Marxist approaches to compile detailed accounts of class groupings and uneven capitalist class relations as they varied over time and space. Edited works such as *Social Relations and Spatial Structures* (Gregory and Urry, 1985) and *Class and Space: The Making of Urban Society* (Thrift and Williams, 1987) illustrate this trend. These works focused on the class patterns and struggles that occurred in different industries (e.g. coal and steel: Cooke, 1985; and professional and management services: Urry, 1987); different places and ethnic contexts (Smith, 1987) and different scales (Walker, 1985). They also concentrated on class differences in housing and the ways class was reproduced and embedded in the urban landscape (Edel, 1982; Forrest and Murie, 1987; Pratt 1989; Rose, 1987). For instance, Pratt was to review the structure of capitalist cities and highlight the specific class segregations that are regularly perpetuated:

> Higher-income residents living in suburban areas have better access to medical facilities and recreational opportunities. They suffer less from the negative externalities of living near noxious facilities, such as air and noise pollution. ... [In contrast] Workers in competitive or secondary sector jobs are trapped in the inner-city rental housing market because they are unable to obtain mortgage credit for home purchase. ... In this way, the effects of labourmarket segmentation have been extended and intensified by residential differentiation and metropolitan fragmentation. (Pratt, 1989: 92)

These works contributed to wider historical and sociological interests in class by showing that 'geography matters' (Massey, 1984). They showed that the economic and social life of individuals and groups would not only be shaped by capitalist modes of production and struggles between different classes. Rather, the organization of production and associated class formations, reproduction and struggle would vary over space, according to different combinations of environmental, economic and socio-cultural conditions, institutions and practices. Marxist-inspired geographies of class thus gave additional place- and region-specific critiques of class differences and conflict.

3.3.2 Documenting class dynamics: studies of class change

As class analyses became well established in social geography through the 1980s, wider Marxist studies were also documenting major crises and changes occurring in capitalist economies and societies throughout the 1970s and 1980s. After the immediate post-Second World War era of modernization and industrial or Fordist eras of capitalism, new patterns were forming. An increasingly global form of capitalism was emerging as new international divisions of labour were established, linking labour relations and production processes in different countries (Frobel et al., 1980; Schoenberger, 1988). Cities, regions and nations were studied for both their locality-specific characteristics and their positions in, and links to, wider circuits of capital and globalized production (Frobel et al., 1980; Herod, 2000; Jones and Womack, 1985; Massey, 1983; Schoenberger, 1988). These developments encouraged greater attention to geographies of class and change.

Although class relations had often been recognized as dynamic (e.g. Cooke, 1985; Rose, 1987), work in the latter 1980s and 1990s paid increasing attention to the changes occurring in the social practices and spatial variations of working and middle classes as well as the increasing importance of particular class factions and new class identities. New tendencies for capital to become increasingly mobile and establish differentiated global production systems meant that class-sensitive analyses of state-driven economic de/re-regulation and industrial restructuring were established. Traditional attention to working-class struggles continued to be highlighted as geographers noted the impact of capitalist restructuring, the movement of investments and altered production arrangements. For instance, studies of de-industrialization and the social impact of industry restructuring, downsizing and job losses recorded the waning political power of working classes and the difficulties facing labour movements (Clark, 1989; Merrifield, 2000). In a similar fashion, Herod (1991a) has traced how 'capital flight and plant closure' in the USA resulted in a class politics based on labour movement concessions and confrontations that varied in both union and local government arenas (see Part V for the wider context that has influenced Herod's work in this area).

More recently, Merrifield (2000) has shown how historically more secure, middle-class workers are also facing a decline in working conditions, wages and job security as corporate interests downsize their labour forces and develop contingent and temporary labour arrangements that provide companies with more flexibility. Merrifield uses Marxist theory to argue:

[T]echnical and white-collar workers retain their privilege and specialized status only insofar as they fulfil the needs of capital. This is a rather precarious privilege, vital at a certain stage of the business cycle yet disposable and superfluous at another stage. ... In America throughout the 1990s, an intermediate stratum – the fabled middle class – is being rapidly absorbed into an ever-expanding mass of 'working class' employment, and its erstwhile badge of honor now reads 'position abolished'. (Merrifield, 2000: 187)

While the power of mobile and flexible capital is acknowledged, Merrifield also documents some of the new social responses. He notes how the interests of traditional blue- and white-collar workers are merging under new basic-wage based class politics such as the 'Living Wage' campaign:

These have proliferated across the country in many different guises, incorporating many different people – white-collar, blue-collar, and black turtle neck – with different skin colors and ethnicities and gender, employed in different kinds of work, who are nevertheless pulling together in unusual and unexpected ways. For all comers, living wage is a quest for economic justice. ... As jobs have been outsourced, made part-time or irregular, as real wages plummet and as the existing minimum wage of $5.15 per hour fails to lift those in work above the poverty threshold, demands for a living or 'family' wage have come from the needy. (Merrifield, 2000: 195)

The second feature of geographies of class change has involved the detailed analyses of specific class fractions and their affect on the social value and use of different spaces. Following Wright's (1985) gradational categorization of class under capitalism, numerous geographers developed British studies of the 'service class' – a fraction or subgroup within the middle class. This reflects economic changes in the structure of advanced capitalism that saw the major growth in white-collar (often salaried) workers employed in public sectors, commercial administrative sectors or social service industries. This class neither owns the means of production nor participates in waged work where their labour contributes directly to commodity production (Thrift, 1987). However, the growth of these jobs and the expansion of their consumer and political power have made them a significant class in contemporary societies such as Britain (Cloke and Thrift, 1987; Thrift, 1987). Thrift explains:

> The service class constitutes an opulent and growing market. Thus there is a ... whole set of distinctive niches for consumer goods and financial services and so on. Manufacturing, service and construction industries have all become more and more geared to the needs of the service class as it has grown. ... [Politically] members of the service class tend to emphasize a degree of stability and order, expressed in issues like their attitude to heritage and the environment [and] quite clearly the service class is a conservative class in its political allegiances. (Thrift, 1987: 225–7)

While Thrift documented some common trends within this particular class, other studies have shown intra-class conflict and geographically specific tensions resulting from these class changes. For instance, Cloke and Thrift (1987, 1990) documented the intra-class disputes developing in rural and commuter areas where middle-class groups have concentrated in southern England. This increased clustering of influential and affluent classes has resulted in the 'consumption of the countryside', where those that can afford it move into rural areas to create and consume picturesque country lifestyles (Cloke et al., 1995; Phillips, 1993; Urry, 1995a). These changes have led to new values of 'rural' space and an altered class politics from one that had linked landowning classes with agricultural labouring classes. In contemporary forms, traditional agrarian work-based politics was replaced as the middle class has established non-agricultural country living. This 'new' class has created lifestyle landscapes (Figure 3.4) and mobilized its own resources and entered disputes over local development, tourism, landscape maintenance, heritage preservation and environmental management (Cloke and Thrift, 1987; Marsden et al., 1993; Murdoch and Marsden, 1994, 1995; Urry, 1995b).

The development of the expanding middle classes, the decline in political uniformity and potency of traditional working-class struggles, and the growth of post-Fordist or post-industrial capitalism has also resulted in a third set of geographies around class change – those based on consumption.

FIGURE 3.4   LIFESTYLE LANDSCAPES OF THE NEW RURAL MIDDLE CLASSES

Source: Panelli, private collection.

While investigations of work-based class inequalities has continued to some extent, findings from the geographies of middle and service classes have led to a consideration of the consumption processes and class conflict arising in this new era of capitalist societies. Geographies of class have continued using Marxist-inspired critiques of class conflict, but other works have given greater attention to cultural processes. Social analyses of housing and gentrification have been key arenas for these developments.

Weberian approaches to class, relating social groups with different life-chances and status, has underpinned early studies of housing where class and consumption processes are linked (Saunders, 1984). Weber's three-fold attention to class, status and political/party interests resonates with many recent studies of gentrification. In these cases middle-class interests in 'liberal ideologies' and their market capacities in terms of income, skills and education are seen to influence changes in housing and class structure in urban settings (Goldthorpe, 1982; Halsey et al., 1980; Ley, 1980, 1987). More recently, Ley's (1994, 1996) work on the reshaping of inner-city spaces has concentrated heavily on intersections between class and culture while building his argument regarding the market and political influence of the new cultural class as a subgroup of the middle class. He notes:

> This rank [of middle class] ... includes professionals in the arts and applied arts, the media, teaching, and social services such as social work, and in other public- and non-profit-sector positions. These cultural and social professionals, who I will refer to as the *cultural new class*, share a vocation to enhance the quality of life in pursuits that are not simply economistic. Their imagineering of an alternative urbanism to suburbanization has helped shape new inner-city environments, where they are to some degree both producer and consumer. (Ley, 1996: 15)

Drawing on Weber's and Bourdieu's attention to the role of artists and aesthetic values, Ley (1996) argues that this new cultural class seeks to form social relations and reconstitute gentrified spaces with attention to aesthetic and pleasurable lives. He explains: '... the new cultural class have not only a heightened sense of the aesthetic, but also "urge pleasure as a duty. This doctrine makes it a failure, a threat to self-esteem, not to 'have fun'"'... Conviviality becomes a style of life' (Ley, 1996, and citing Bourdieu, 1984: 367). This type of emphasis on culture and consumption has also influenced social geographies of contrasting classes, as illustrated by May's study of tensions between traditional working classes and in-migrating cultural classes in inner London. As shown in Box 3.2, the incomers contribute to a reworking of the social identity and streetscape of Stoke Newington; traditional shops and pubs are replaced or renamed and experiences of place are redrawn starkly along class lines. Working class residents express a sense of decline and loss while new cultural class in-comers value the English history of the neighbourhood, the Victorian architecture, old parish church, and Clissold Park – complete with up-market café.

Box 3.2 Class change and gentrification in Stoke Newington, London

Stoke Newington – a long-time English working-class neighbourhood - has recently seen many changes. In-migration has challenged both ethnicity and class-based identities.[1] In terms of class, significant numbers of new cultural class residents are moving into the area and influencing the social relations and sense of place. They seek the 'Englishness' of the 'Stoke Tup' and the gas lamps of Stoke Newington Church Street leading up to the local parish church. They are also creating demand for different economic activity, including 'exotic' restaurants and organic food.

The StokeTup, previously The Red Lion and The Magpie and Stump

Stoke Newington Church Street complete with gas lamps and church spire

Blue Lobster Restaurant

Fresh and Wild: organic food shop

CHANGES:

In the place of the old barbers there is now a kite shop; instead of the butchers, a delicatessen. The fish and chip shop has long gone, replaced by ... (reassuringly expensive) [Asian] restaurant[s] ... in recent years shop fronts have been restored and the names of pubs changed to something more in keeping with the area's literary aspirations (*The Three Crowns* to *The Samuel Beckett* for example) and in 1986 Hackney Council spent some £27,000 installing a set of 'traditional English gas lamps' along the length of Church Street. In 1982, both that street and the adjoining streets of Shakespeare Walk and Milton Grove were granted conservation status as the area's new residents demanded the geographical extension of history. (May 1996: 197, 202)

(Continued)

Box 3.2 (Continued)

CLASS DIFFERENCES

[F]or these [new cultural class] residents the area is being re-imagined as the quintessential English village and this secures a rather ambiguous relationship between the two groups. Whilst the working-class residents welcome the arrival of recent gentrifiers as a means of bolstering the area's Englishness, the sense of Englishness around which this new image is being constructed is also working to destroy working-class understandings of the neighbour-hood. (1996: 197)

[1]For the purposes of this chapter, note is made of Stoke Newington's class changes and May's reading of class, however, racialized tensions are also addressed in his work.

Source of photographs: J. and V. May and C. Hall, private collection.

This type of geography contrasts with the radical and critical emphases of Marxist approaches. For instance, Neil Smith's (1992a, 1993, 1996) works have highlighted the uneven capital investment processes and frequently violent political struggles that lie behind the new constructions of gentrified consumption landscapes in many cities. Smith's account of gentrification in inner-city New York includes analysis of how class changes in the Lower East Side saw struggles between city administrators and police supporting the investment in gentrified residential areas at the expense of homeless, evictees and squatters who wished to 'take back the parks', including Tompkins Square Park (see Box 3.3).

In a similar vein, Bridge (1995: 245) argues that current interest in working and middle-class differences in gentrification 'masks a deeper two-class relation (between capital and labour)'. He suggests that the gentrified neighbourhood and the gentrification process are secondary to, and post-date, the fundamental class constitutive relations dividing labour and the class socialization that produces different lifestyle, taste and housing patterns so frequently read in gentrification studies.

Throughout the last four decades studies of class change have tackled the diverse processes and dimensions of class structure and class relations. From the first works studying class struggle over industrial restructuring through to the newer foci on class-based consumption, geographers have acknowledged that differences and uneven experiences occur in space and are influenced by numerous social contexts. Consideration of these complexities is taken further in new critical reconceptualizations of class as a multiply constituted dynamic phenomenon. Studies using an overdetermined approach thus form the third cluster of geographies of class.

Box 3.3 Timeline of a class struggle: Tompkins Square Park, New York

Neil Smith has completed many Marxist analyses of class. During the 1990s he has written significant pieces on both urban middle-class gentrification and homelessness 'under classes'. Taking the case of Tompkins Square Park in New York's Lower East Side, he documents the class processes and social (and political) struggles that occur over the access to - and uses of - a 'public space'.

   The story of Tompkins Square Park (Smith 1993) illustrates the material and spatial conditions that differ between middle classes (who see the park as an urban resource for controlled use and enjoyment by the gentrifying class) and homeless classes (who see the park as a respite and a shelter and a site for housing activism and protest). The following time line is based on events Smith has analysed.

1850s onwards	The park becomes a key site for political protest.
1874	First police riot in the park following a march by the unemployed.
1880s–1980s	Around the park a heterogeneous working-class neighbourhood develops with an alternative culture to white middle-class interests.
1970s–80s	Around the park, gentrification processes develop and housing is redeveloped for middle-class incomers.
1980s	The park becomes a focus for evictees and squatters.
1988	Police attempt to invoke historic curfew to re-evict evictees from the park. Major riot follows (August). Cultural and legal debate over the function and access to this park as a public space.
1988–89	Squats are re-established in nearby buildings and up to 250 evictees are living in the park. Political rallies and events in support of homeless people and against gentrification become a regular event.
1989 (July–Dec.)	Police seal off the park and destroy evictees' shelters (July). Evictees return. Major re-eviction of park dwellers and destruction of shelters. (December).
1991	Pro-gentrification lobby seek wider media and administrative support. 200 park dwellers are evicted (June) and the park is 'secured' with a high fence and 'reconstructed'.
1992	Reconstructed park is re-opened and becomes renewed focus of class struggle.

3.3.3 Class as an overdetermined and multiply constituted process

By the mid-1990s the literature on class geographies began to register the affect of postmodernism and poststructuralist thought as it had developed in human geography more widely. In geography, these two movements focused on objects (e.g. texts, landscapes, postmodern cities, post-rural spaces) as well as the methods or practices of inquiry/research and knowledge construction. Major critiques of essentializing metanarratives such as Marxism had a direct impact on studies of class. For example, see the debate around Marxist essentialism, the privileging of class as a foundational social category, and the calls for anti-essentialist approaches (Gibson and Graham, 1992; Graham, 1992; Laclau and Mouffe, 1985; Peet 1992; Resnick and Wolff, 1987). In an equally critical trend, methods of researching and reporting geographic knowledge/'Truth' on class (structure, formation, exploitation or struggle) also came under close scrutiny.

Illustrative of this type of critique, Katherine Gibson and Julie Graham's work has received widespread recognition (and debate) as they have worked individually from Marxist backgrounds to publish frequently as a single persona – J.K. Gibson-Graham – exploring poststructural practices in re-presentation and authorship alongside their critique of capitalism and Marxist geography. Initially they each highlighted the greater complexity of class issues than was frequently acknowledged in Marxist geographies (Gibson and Graham, 1992; Graham, 1992). They have shown that both feudal and capitalist class processes complicate social relations for men and women in the Australian mining industry via the gendered division of paid 'productive' work and unpaid 'reproductive' work in company-owned housing that is distributed according to men's status as single or partnered (with one or more 'dependents'). Later, experimenting with a single collaborative author-identity, Gibson-Graham (1996a) further developed a critique of the concept of capitalism (and class structure). They argued that Marxist theory approached capitalism as an essential, heroic, privileged, master concept that explains all economic conditions and social relations. They deconstructed this privileged view of capitalism and sought to show a variety of relations and modes of production could exist simultaneously. For instance, again in reference to coal mining they argued:

> On the Central Queensland coalfields an important condition of existence of capitalist mining operations is the feudal household. ... Usually only capitalist class processes associated with the coal industry are recognized as structuring life and struggles in mining towns. We have argued for the representation of feudal class processes associated with household production as another important axis of social organization and struggle. Men's participation in both capitalist and feudal domestic class processes, and women's almost exclusive participation in feudal ones, overdetermines the political roles women are able to take in mining communities. (Gibson-Graham, 1996a: 233)

Building on from early debates between Graham (1992) and Peet (1992), they have promoted ***anti-essentialist*** poststructural readings of capitalism and class as culturally constructed in material and discursive ways. This has been a major point of departure from the mostly 'economistic' focus of previous class analysis, making greater space for social and cultural dimensions. Arguing for an *overdetermined* approach, Gibson-Graham see class as an overdetermined process of producing and appropriating surplus labour:

> We understand class processes as overdetermined, or constituted, by every other aspect of social life ... class is constituted at the intersection of all social dimensions or processes – economic, political, cultural, natural – and class processes themselves participate in constituting these other dimensions of social existence. This mutual constitution of social processes generates an unending sequence of surprises and contradictions ... class and other aspects of society are seen as existing in change and as continually undergoing novel and contradictory transformations. ... An overdeterminist class analysis examines some of the ways in which class processes participate in constituting and, in turn, are constituted by other social and natural processes. (Gibson-Graham, 1996a: 55).

Following this line, Gibson-Graham (1995, 1996a) produced various readings of men's and women's multiple positions and economic, social and political experiences in the mining industry (see Box 3.4). These types of study show how economic processes, gender, ethnicity, regionally-specific political culture and history each affect and constitute the classed lives of individuals, households and class struggle.

3.4 CONCLUSIONS, CHALLENGES AND CONNECTIONS

This chapter has sketched the core traditional approaches to class, based on Weberian and Marxist theories. We have seen how geographers concentrated first on the analysis of class groups and relations and how these were played out spatially in differing ways. Continuing the notion of class relations as dynamic, growing attention to class change has resulted in both urban and rural geographies of class as a frequently altering and contested process. Contemporary geographies of class draw on the long and rich critical tradition, as shown in the preceding account of Marxist approaches. However, the current outlook involves geographers in moving beyond relatively exclusive, single-focused work on class and taking into consideration other complexities and interconnections. Whether explicitly following the Gibson-Graham style of poststructuralist inquiry or a modified critical Marxist approach such as Smith's (2000), geographers are turning their focus to the implications of considering class within wider cultural and political bases of difference. Now, class is being addressed as a complex process and as one – among a number – of social categories that involve multiple cultural and material dimensions. This situation is one of challenges for traditional class specialists. But it also opens up studies of class to new and potentially productive connections with other social geographies. Three directions are worth noting.

Box 3.4 Overdetermined geographies of class: reading mutual constitution

Gibson-Graham (1996a) argue a new conceptualization of class:

[W]e define class simply as the social *process* of producing and appropriating surplus labor (more commonly known as *exploitation*) and the associated process of surplus labor distribution. … We understand class processes as overdetermined, or constituted, by every other aspect of social life. (1996a: 52, 55)

Gibson-Graham illustrate their account with the case of 'Bill' and 'Sue', a couple living in and beyond a mining society.

Bill's class location is difficult to ascertain. He is a wage laborer from whom surplus value is derived. He has little control over his own labor process. Yet he owns shares in productive capitalist enterprises and receives a small share in the profits made by the mining company that employs him. He is a member of two political organizations with quite antithetical philosophies and is active in both… Bill's membership in the 'working class' can only be secured by emphasizing some of the relations in which he participates and de-emphasizing others. …

If we emphasize Bill's role in the construction of Sue's class identity, she might be seen as a member of the working class by virtue of her marriage. …Before she moved to the mining town, however, she was employed as a nurse in a supervisory position that distinguished her from those with no control over their labor or that of others. And before she migrated to Australia she was the relatively well-off daughter of a member of the petit bourgeoisie in the Philippines. Given that one of these 'class locations' belongs to her husband, one to her past, and one to her father, class as a social category would not seem directly relevant. (1996a, 61–2)

Rather than pursue an analysis of class categories, Gibson-Graham describe the complex class processes and interconnecting aspects of work, family and community life for this couple. The following diagram highlights the key aspects and links that shape the class processes for these two individuals.

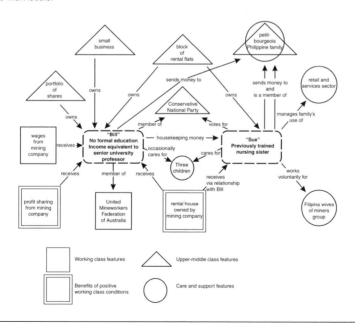

First, the influence of postmodern and poststructural geographies is being felt widely. Geographers are 'reading' class and attending to the 'discourses' that play a part in class formation and politics. In terms of class formation and the way different class identities are understood, work such as Jarosz and Lawson's (2002) investigation of economic restructuring and rural 'rednecks' in the USA illustrates how discourses generate and circulate meanings that link class and ethnicity in various settings. Other complementary geographies consider the 'landscapes' that contemporary economies and societies produce in terms of class cultures and spatial patterns. For instance, Herod's recent work analysing the 'economic landscape' of containerized ports in the USA illustrates how different class struggles are taken up in two trade unions via a struggle over discourse and the meanings of work traditions. He reports that:

> Local discursive regimes were ... powerful influences on the state and courts and helped shape the geography of work in the industry. The Supreme Court's decisions ruling in favour of dockers rested to a greater extent on the ILA's successful discursive representation of work traditions prior to containerization than it necessarily did on the patterns of in situ job loss. ... Only through its ability to construct a more convincing – if not necessarily 'truer' in some absolute sense – representation of patterns of work in the industry was the ILA able to retain work at the piers for its members. (Herod, 1998a: 187)

This type of work illustrates how some of the discursive foci and tools employed in poststructural work can complement and enhance traditional topics in class geographies such as class struggles. Herod shows that class conflict need no longer be analysed as a discrete struggle over particular wages and work conditions, but is now 'read' as a struggle that is both materially and discursively constructed and played out in many places and at many scales (from wharves and piers to judicial courts – for more details see Part V). These findings link class to wider notions of identity and power (that are raised in Chapters 7 and 8), for accounts of class such as Herod's suggest the importance of ideas of class identity and the power relations of class struggle or conflict (see also Wills, 2001).

Second, affected by geographies of ethnicity and gender, studies of class are being written that engage with equally important social relations and identities based on gender and ethnicity (to be explored in Chapters 4 and 5). These include investigations into intersecting inequalities that form around axes of class, gender and ethnicity (England and Stiell, 1997; McDowell, 2000; Pratt, 1997, 1999a). And while there have been some fears that a diversification of class analyses may dilute and fragment the critical power of class analysis, McDowell points out that newly formulated critical geographies of class are never more necessary:

As capital becomes globally mobile, exploiting a working class that is divided by race and gender in geographically-diverse ways, as work is casualized, trade union movements defeated and neo-liberal social politics are implemented, it is theoretically and also politically crucial that geographers explore both the diverse dimensions of the new gender and class order of the twenty-first century as well as ways in which this order may be challenged and made more equitable. (McDowell, 2000: 461)

A third challenge and opportunity for geographies of class involves elaborations of an **anti-essentialist** and **overdetermined** approach to class. The heterogeneity and interconnected qualities of class in overdetermined analyses complement current discussions on the global and dynamic nature of post-industrial capitalism as local patterns and differences continue to influence how class is played out in individual settings or specific industries and policy arenas. Katherine Gibson's (1998) work on 'class and multiplicity' urges an engagement with diverse class experiences while also pointing to the need to extend these nuances to challenge the economic dominance of many analyses:

[C]lass is still constituted as a coherent and unified aspect of one's identity. But, in fact, ... it is obvious that we all perform in many different sorts of class processes within one day, as well as over our lifetime. The various mix of class processes we are involved in may include the capitalist class process of producing surplus value and having it appropriated, or distributing already appropriated surplus value. But it could just as easily include other modes of surplus labor production, appropriation, and distribution. Many of the other economic transactions we engage in may, in fact, not involve class transfers but may be overdetermined by other social relations involving power and oppression. [However, u]ntil the productive dis-ease that has been introduced into our discussions of gender or race by difference theorists is introduced into our language of class, the economic diversity that is as much an aspect of difference as anything else will remain suppressed both in knowledge and in policy practice. (K. Gibson, 1998: 309–10, 312–13)

Finally, within Marxist circles, attention to class is being 'invigorated' by the 'salutary' poststructural critiques suggesting that Marxist approaches are too 'modern' in theory and practice (Castree, 1999). Re-readings of Marx are now being conducted with appreciation for the challenges being raised by these contemporary critiques. Castree (1999: 141) has argued for the possibility of capitalism as an open system including 'non-capitalist' elements, and that a geography of class can be developed that 'retains the Marxian claim that it is a key axis of social inequality as well as a central vehicle for political organization, yet without closing it off to other forms of social identity'. Class may now be acknowledged as a 'domain of social heterogeneity not unity'. But heterogeneity does not dilute the inequalities that have spurred so many social geographers' work. Indeed, Castree (1999: 153) points out that the notion of class as 'positionality' within global patterns of capitalism is increasingly pivotal to understanding political possibilities for the redistribution of wealth, and the possibility of 'fashion[ing] a global political movement against capitalism'.

In conclusion, while geographies of class have produced a large corpus of primarily critical, Marxist-inspired accounts of unequal class relations, recent challenges have opened up ways of connecting (and theorizing) class in connection with other social axes. Questions of gender, ethnicity and sexuality (that follow in the next chapters) also intersect with class. Across these multiple differences, class-based spatial and social inequalities within capitalism continue to stimulate geographers (Castree, 1999; Herod, 1998a; Smith, 2000; Wills, 2001). Thus, studies of class (based around surveys like Box 3.1 at the beginning of this chapter) will continue to point to further directions in social difference, individual and group identity (Chapter 7) and the power relations and struggles (Chapter 8) with which we occasionally – or regularly – engage.

SUMMARY

- Class has been a core category of social difference for many years, enabling geographers to describe and study social and economic inequalities between different groups of people.
- Theoretically, *gradational* descriptions of class can be contrasted with *relational* explanations and poststructural readings of class-*processes.*
- Class has most frequently highlighted people's material experiences as related to their position in the economic system operating in their society, including their income, material conditions of life (such as housing), and their employment position and experience of work.
- Weberian theory explained class as an expression of individuals' life-chances in relation to resources and returns they could access in a market-based society.
- Weberian theories of class have stimulated studies of class in relation to people's position in labour and housing markets and the links between class, social status and political or party interests.
- Marxist theories of class have emphasized the unequal class relations and struggles occurring within capitalist modes of production.
- Marxists geographies of class have shown how capitalist class relations vary, and continually reshape, social relations across space in dynamic and uneven ways.
- Attention to conflict through capitalist restructuring and class fractions have continued to stimulate Marxist-inspired critical class analyses.

- More recently, geographies of class culture have increased in importance, highlighting the way processes of consumption are equally important in shaping spaces and social relations.
- Poststructural theories of class have suggested class should be read as an overdetermined process (mutually connected to other dimensions) rather than a core category.
- Geographies of class increasingly note connections with other categories of difference (e.g. gender, ethnicity, sexuality) while retaining an ongoing interest in the continuing inequalities and exploitation of class processes.

Suggested reading

A useful introduction to the concept of class is provided in Edgell's (1993) very readable book. An accessible analysis of Marxist-inspired class *fractions* is provided in a paper by Cloke and Thrift (1987) when they draw on Wright's typology to study the intra-class tensions within rural middle-class groupings. In terms of class change, the edited book by Butler and Savage (1995) provides a range of perspectives, and Smith's (1996) book and May's (1996) paper give detailed accounts of urban class change in New York and London respectively. Finally, important developments in geographies of class include Herod's book on *Labor Geographies: Workers and the Landscapes of Capitalism* (2001a) and Gibson-Graham (1996a) and K. Gibson (1998) chapters on class as an overdetermined process.

Gender

4.1 INTRODUCTION

'Ben': Thinking about gender makes you realize all the assumptions you make, ... we're acting and expecting things based on gender. ... In my [student] flat, the guys got the lease, then we looked round for some girls 'cuz we knew the place would work better with them, y'know ... [laughter]. That's sexist, but it's *true*! They make sure we all share the cleaning and stuff more – otherwise we'd just leave it ...

'Grant': But then you can't expect too much 'cuz like they [female students] expect you to do stuff too, round the flat. Then they'd like you to walk them home at night, ... and if you're watching the rugby on TV ... they get pissed off. ... [So] it's everywhere, gender: your actions as a guy, and their expectations as well.

(Excerpt from class discussion, University of Otago, 2000, original emphasis)

Following the influence of feminism and the changing status of women in western societies over the last 50 years, gender issues are more frequently acknowledged throughout geography. This has included tracking the effect of gender throughout the history and practice of geography as a discipline (Blunt and Rose, 1994; Maddrell, 1998; Monk and Hanson, 1982; Phillips, 1997; Rose, G., 1993). It has also stimulated the growth of gender-related academic networks, and resulted in geography students' willingness to consider gender issues in their everyday lives. As noted by the male students quoted above, gender forms one of the major axes of social difference in our lives. It influences the domestic and social arrangements we make and it affects the assumptions and expectations we hold.

In each of these academic and personal situations gender is distinguished from sex. Until recently **sex** has been understood in terms of our biological characteristics while gender has been seen as a socially-constructed phenomenon expressed through ideas of **masculinity** and femininity. Gender also affects our experience of material life, cultural practices and power relations. Geographers – from students to professors – can all recognize the micro-geographies of gendered places in their everyday lives (e.g. student flats and work tea-rooms) and the gendering of physical and imagined spaces in wider society (e.g. building sites and TV soap operas).

This chapter considers how gender has been addressed theoretically (section 4.2) and how geographers have investigated gender over the past three decades (section 4.3). This latter section mirrors theoretical developments occurring in the discipline and includes a discussion of socio-material geographies of gender (focusing on the gender relations shaping work, living arrangements and space), and socio-cultural geographies of gender (focusing on the construction and negotiation of gendered identities, differences and politics). Finally, in section 4.4, the chapter closes by highlighting the challenges and wider connections that currently face geographies of gender as scholars engage with recent geographic foci and developments in wider social theory. To complement this structure, a biography in Part V covers many of the same themes from the perspective of a single scholar, Jo Little. Little's work reflects the range of theoretical approaches that have been taken in gender studies as well as the way socio-material and socio-cultural foci have been developed in rural examples of research.

4.2 DEFINING GENDER: DRAWING ON (DISCIPLINARY) EXPERIENCE AND THEORY

As noted in Chapter 2, the first radical initiatives to focus on gender were influenced by feminism and reflected the radical turn in the whole discipline of human geography. Throughout the 1970s and 1980s radical social geographers came to question variations in the material conditions of life, experiences of work, access to services and participation in different sectors of society. They demonstrated inequalities by mapping and analysing uneven spatial patterns and conditions. They sought to explain these conditions by analysing the power relations affecting the social and economic organization of life at various scales (e.g. households, neighbourhoods, cities, regions and nations). Such explanations drew on critical Marxist and feminist theories.

Feminists working in this climate began to critique geography as a male-dominated discipline. They sought to write women into the academic environment and the substance of geographic inquiries. For instance, British feminist geographers were to argue:

> In common with other approaches in geography that are critical of mainstream work, we must analyse and understand why women remain in a subordinate position. ... We are concerned to introduce the idea of a *feminist geography* – a geography which explicitly takes into account the socially created gender structure of society; and in which a commitment

both towards the alleviation of gender inequality in the short term and towards its removal, through social change towards real equality, in the longer term, is expressed. (Women and Geography Study Group, 1984: 20, original emphasis)

Many geographies of gender were to follow throughout the 1980s, each seeking to distinguish between the experiences and actions of men and women in different contexts. Two fields of endeavour developed. First, studies documenting the inferior position of women in the discipline were established (McDowell, 1990; Momsen, 1980; Monk and Hanson, 1982), and second, feminists worked to theorize gender more broadly.

Feminist theoretical writings aimed to develop detailed conceptual frameworks for explaining how different experiences and inequalities occur – and are maintained – between men and women. During the 1980s feminist scholars debated how best to analyse these material, social and political differences. Theorists were particularly interested in ways to *explain* (not just describe) the disadvantaged position of women in many areas of life (Bowlby et al., 1989; Women and Geography Study Group, 1984). Early explanations of difference between men and women rested on sex- and gender-role theories. These types of role-theory postulated that men and women would learn and operate within certain social roles. These roles were understood to be divided by expectations of men and women that ranged from types of work, to psychological characteristics and interests in economic and political matters. Examples of this early approach include Tivers' (1978) discussion of women's role in society and the need for geography to critique the inequalities following from 'sex-role differences'.

During the 1980s, a move from role theory to the consideration of **gender relations** was a significant development. Gender roles were considered a theoretically static and limited form of explanation compared to ideas of gender relations:

> [G]ender roles [focus on] ... a relatively static set of assumptions about the supposed characteristics of women and men, rather than an active social process of gender relations through which male power over women is established and maintained. ... We argue that it is through gender relations that gender roles ... are formed and reformed, and gender experienced, in particular places, by women and men. (Bowlby et al., 1986: 328)

Gender relations theory focused on the dynamic social processes that ensure male power and priorities over women. Adopting these ideas of gendered power relations enabled social geographers to explore the bases and practices of power relations that unevenly affected men's and women's experiences and access to different arenas and spaces. It also stimulated a great deal of debate. One key exchange occurred in *Antipode* in the later 1980s. In this case, Foord and Gregson (1986) argued that just as capitalism is a specific mode of production, so too, **patriarchy** is a specific form of gender relations that disadvantage women (Bowlby et al., 1986). Debate followed over whether patriarchy could be seen as equal with, or preceding, capitalism in terms of its capacity to generate inequality and exploitation. Foord and Gregson had presented an agenda that privileged patriarchy, and gender relations more broadly, as a core focus for feminist research. In contrast, McDowell (1986) argued that patriarchy and the oppression of women had a fundamental material base, and therefore it was women's positions within capitalism that explained unequal gendered experiences. For example, she noted: 'it is rather the particular way in which working-class women undertake the domestic labour necessary for capitalist social reproduction that is the key element in their oppression' (McDowell, 1986: 319). Further to this debate, Johnson (1987) questioned the dominance of economic considerations, suggesting that consideration should be given to the 'necessity' or naturalization of patriarchal and heterosexual relations that could be interpreted in such accounts. This heralded the social and cultural geographic literature that was to develop more fully in the coming decade around the issues of gender and sexuality (see Chapter 6).

The stimulus of this debate supported many socio-material geographies of gender (see section 4.3.1), focusing on gender differences in experiences of work, living arrangements and space – with more or less emphasis on patriarchal, capitalist or heterosexual power relations. While the inequalities between women and men continued to be analysed as both an economic and social geography project (e.g. Pratt and Hanson, 1988), theories of gender in the 1990s gradually reflected the fact that, even when some women moved closer to economic equality with some men, broader social and cultural beliefs and practices still influenced the opportunities and expectations of men and women. Drawing on the 'cultural turn' and developments in postmodern, poststructural and postcolonial theories, feminist geographers theorized gender based on notions of gender *differences* and gender *identities* (e.g. McDowell, 1999; Pratt, 1993; Pratt and Hanson, 1994). This theoretical approach recognizes societies and cultures as gendered, not in a fixed distinction between homogeneous categories of men and women, but in diversely constructed and contested ways. Gender, including notions of masculinity and femininity, is recognized as constructed through **gender identities** that may associate men and women with certain spaces, industries, activities, appearances, interests and behaviours. Using poststructural interests in the importance (and instability) of language and meaning, gender identities have been read through studies of landscapes, language, bodies and broad discourses circulating in workplaces, media, political systems and so forth (Agg and Phillips, 1998; Duncan, 1996; Liepins, 1998a; Longhurst, 1995; Longhurst and Johnston, 1998; McDowell, 1995; Mackenzie, 1994). Details of this type of work are provided in section 4.3.2.

Equally important in these recent developments has been the recognition that gender is a contested and differentiated category. Feminist geography has been challenged and stimulated by acknowledgements of the diversity and differences that divide the category of 'woman', and the increasing interest accorded to concepts of masculinity. In both cases, geographers have increasingly paid attention to diversity and the way gender identities are performed and contested. In the former case, attention to gender differences between women and between feminist practices of geography has been important (e.g. Jones et al., 1997; Pratt, 1999b; Pratt and Hanson, 1994). In the latter case (to be discussed in section 4.3.3), geographies of gender have been expanded by a growing interest in the construction and performance of masculinities (e.g. Jackson, 1991, 1994; McDowell, 2002a, 2002b; Massey, 1996; Woodward, 2000).

4.3 CONTRASTING GEOGRAPHIES OF GENDER

Geographies of gender can be recognized in at least three broad clusters: work focusing on women in geography; socio-material geographies of gender; and socio-cultural geographies of gender. These clusters form the structure of the following discussion and are summarized in Table 4.1. It should be noted, however, that the practice and methods of conducting and writing geography have also been an important gender project. Although this work spreads beyond social geography, a wide range of feminist and gender-sensitive methodological literature has been established. This has focused on geographic practice through reflections on how research is designed, how data is gathered and how analysis is presented (McDowell, 1992; Mackenzie, 1984). Techniques of quantitative and qualitative research have been reviewed, as have the conditions under which researchers position themselves in relation to their research and written results (see Pratt, 1993; and special issues of *The Canadian Geographer*, 1993, and *The Professional Geographer*, 1994). Following wider feminist critiques of social science in general, these works demonstrate reflections on the techniques and power relations inherent in the research process (Herod, 1993; McDowell, 1992; and Rose, G., 1993). This type of work has, questioned the ethics and politics involved in conducting research, provided reflections on the partial and situated positions of both researchers and the researched, and suggested possible alternative approaches to research, including more inclusive, participatory and action-oriented studies (Gibson-Graham, 1996b; Jones et al., 1997). Beyond these broad discussions, however, the three clusters of substantive gender geography have developed a pace. While Table 4.1 and the following sub-sections treat these as largely separate initiatives, authors in fact have merged, and worked across, these foci.

TABLE 4.1   THREE CONTRASTING GEOGRAPHIES OF GENDER

Chronological geographies of gender	Type	Description
First (1970s–90s)	Women in geography	Account of unequal experiences and opportunities for men and women within the discipline of geography.
Second (1980s–ongoing)	Socio-material geographies of gender	Primarily socialist and radical feminist accounts of the unequal gender relations and socio-material conditions experienced by women (and to a lesser extent men).
Third (1990s–ongoing)	Socio-cultural geographies of gender	Primarily poststructural accounts of *both* masculinity and femininity associated with constructed and contested gender differences and identities.

4.3.1 Women in geography

Initially, gendered geographies emerged across the discipline as feminist academics drew on the momentum from the 'second wave' of feminism that developed in western societies during the 1960s and 1970s (Blunt and Wills, 2000). Geography-specific agendas mirroring this 'Women's Liberation Movement' included attempts to increase the recognition of women and their unequal experiences within the discipline. The first feminist publications appeared in *Antipode* in the early 1970s, exploring the position of women within geographies of urban space (Bruegel, 1973; Burnett, 1973; Hayford, 1974). By the end of the decade momentum was building. Feminists moved into the 1980s questioning both the position of women within the discipline and the types of geography that were being constructed (McDowell, 1979; Mackenzie, 1984; Mackenzie et al., 1980; Momsen, 1980; Monk and Hanson, 1982; Tivers, 1978; Zelinsky et al., 1982).

Feminist social geography reflected on the status of women both as subjects of geographic research and as professionals within geography circles. In early work, Hayford argued that geography was influenced by broader sexism within society: 'Those institutions that express, or seem to express, a dominant male role are taken as expressive of the whole order of society, while women are assumed either to have no distinct role or to be in a state of continuous adjustment to this male-dominated and male determined order' (Hayford, 1974: 2). Complementing this position, some geographers launched a wide-ranging review of women within the profession of geography. Results showed that, when compared with men, women in the discipline were overwhelmingly positioned at lower scales with less secure conditions, and more expectations about their support and ancillary tasks. These works concluded that women geographers faced more difficult conditions than their male colleagues and that some deeply unequal relations and practices existed within the profession (McDowell, 1979; Momsen, 1980; Monk and Hanson, 1982).

Feminist approaches were also building a critique of the content of social geography. For instance, Tivers noted the gross assumptions being made by conventional (male) geographers when studying behaviour and activity patterns. For example, she argued: 'we cannot fully understand consumer behaviour without considering the specific role which women play in urban society' (1978: 303). Drawing on the then-current sex-role theory, these 'roles' and assumptions were argued to be operating throughout the writing and practice of geography. Cumulatively, feminists realized that geography was being limited by many unquestioned social arrangements and assumptions. In terms of the coverage of geography, Monk and Hanson (1982) argued that geographers had tended to exclude half of the subjects that ought to be considered within *human* geography. They criticized conventional geography for assuming that the experiences, behaviours and issues important to men (especially white, middle-class, able-bodied men) would be the norm.

Feminist geographers went on to call for the inclusion of women and women's issues in the whole range of human geographic studies. Pertinent topics for study were also identified in a number of published 'agendas'. These included the need to distinguish between men's and women's experiences, behaviours and preferences (Tivers, 1978) and the need to identify the gendered nature of the spatial division of labour across different economic industries and regions (McDowell and Massey, 1984). Later, Hanson (1992) was to argue that common agendas could be shared by feminists and other geographers, including the study of 'everyday life', the consideration of context for understanding social relations and processes, and the investigation of how meanings about difference are constructed (and even possibly challenged). This early activity – in noting the absence of women in geography and calling for a geography of women – established a base from which a wide-ranging geography of gender has flourished. Influenced by wider trends in the discipline, focus was first placed upon the social and material conditions of women (section 4.3.2). Later attention moved to the diverse cultural constructions of gender *per se*, including the different ways both femininities and masculinities have been developed, negotiated and in some cases even reconstructed (section 4.3.3).

4.3.2   Socio-material geographies of gender: negotiating work, living arrangements and space

The impact of the 'women in geography' critique encouraged many feminist geographers (and some of their colleagues) to investigate the unequal conditions of men's and women's lives. This was a socio-material pro-gramme for it developed in the 1980s when quantitative interests were being widely challenged by radical geographies that focused on the mater-ial inequalities and injustices between and within classes, economies and societies. The resultant geographies of gender reflected this interest in inequalities and they reported divisions of work and life and the gendered organization of space. One example of this type of approach is included in Part V where the early work of feminist geographer, Jo Little, is highlighted, especially her work on women's position and experience in rural societies and local communities.

GENDER DIVISIONS OF WORK AND LIVING ARRANGEMENTS

Following on from the new energies provided by welfare and Marxist geo-graphies of labour relations and class conflict, early gendered geographies sought to 'add women into' the geographies of work, housing and services. For instance, Tivers' (1985) research had used established analyses of spatial activity and movement to show how women's activity spaces were affected by their domestic and paid work and their need to combine travel arrangements for employment and children's needs. Along with other welfare geographies of the time, this type of geography challenged and expanded the traditional spatial models of industrial location and trans-portation that had been based on ubiquitous assumptions about 'rational economic man'. It was now possible to consider the specific needs of different populations (including women).

Using more radical theoretical foci, other scholars began to 'add women into' increasingly popular Marxist geographies of work, housing and the spatial economies of western societies. In terms of employment, McDowell and Massey's (1984) historical investigation of work in English coal, cotton, textile and agricultural industries showed how women's par-ticipation in paid 'productive' work was also unevenly arranged and con-ducted. Production became separated from domestic work and spaces, as national and international capitalist processes changed. McDowell and Massey (1984: 128) argued that patriarchy resulted in locally specific gender relations in order to maintain material gender differences: 'the con-trasting forms of economic development in different parts of the country presented distinct conditions for the maintenance of male dominance ... capitalism presented patriarchy with different challenges in different parts of the country'. Together with Mackenzie and Rose (1983) and Hanson and Pratt (1988), McDowell and Massey's work constitutes a record of how men's and women's employment and economic opportunities have been divided throughout different phases of capitalism. Beyond paid labour, these authors also argued that the rise of industrial capitalism separated 'production for exchange' from 'domestic work' and increasingly devalued the latter and coded it 'feminine'.

Initial works focused on incorporating analyses of women's work into geographic literature, but over time it was also complemented by gender analyses of the variety of men's and women's paid, community and domestic work (Johnson, 1991; Little and Austin, 1996; Milroy and Wismer, 1994; Pratt and Hanson, 1988, 1994; Whatmore, 1991). Studies of both urban and rural employment showed that women were often positioned in inferior work conditions because of their multiple commitments and their lack of access to capital, services or decision-making. For instance, Whatmore's study of British farms showed that expectations about women as wives and mothers, together with their unequal position in terms of farm work, farm management and farm ownership, affected farm women both economically and personally.

> [Women] lack control over the means of production and the products of their labour. These patriarchal labour relations are structured through the interlocking of conjugal kinship relations ... and patrilineal kinship practices which organize the ownership and transfer of property rights. ... Women's rights in land and business assets and involvement in the business decision-making process are principally structured by their *specific* gender position as 'wives'. (Whatmore, 1991: 84)

In a complementary but quite different study, Johnson's account of the Australian textile industry showed the way gender divisions in labour occurred both within the mill and the working-class homes. Box 4.1 illustrates how men's and women's reasons for paid work and participation in domestic work fell into traditional patriarchal groupings. Added to this, gender divisions within the textile mills also maintained men's participation and acknowledgement in more (physically or managerially) powerful and higher status work. Women, on the other hand, worked in 'support' positions in the quieter and 'feminized' spaces such as administrative offices and sewing rooms – or worked as 'minders' of the large winding and weaving machinery.

Cumulatively, these studies of paid, community and domestic work document several key findings, including the following:

- The structure of occupations involves gender divisions of tasks, responsibilities, opportunities and income (e.g. in agricultural and textile industries) (Johnson, 1991; Whatmore, 1991).
- Women's work shapes their experience of other aspects of their lives, including relationships with kin, living and housing arrangements, experiences of travel (transport needs), and interaction with service and state agencies (Fincher, 1991; Hanson and Pratt, 1995; Whatmore, 1991).

- The participation in paid and unpaid work includes negotiating meanings of work and authority within both households and places of paid employment.

Box 4.1 Gender divisions in an Australian textile industry: work, meanings and spaces

Louise Johnson's study of men and women working in a Geelong textile mill involved in-depth fieldwork during the 1980s. Through a range of interviews, observations and photography, she made the following conclusions:

Intersection between paid and home work

Women:
- Have work in the home as an alternative to paid work

- Complete domestic work involving: child care, cooking, cleaning and washing
- Both choose to, and have to work in paid employment

- Earn wages that are crucial but secondary to the household
- Enter working classes as 'interlopers'

Men:
- See no alternative work to paid employment in the public sphere

- Contribute to domestic work and arrangements via their pay and care of the garden and car
- Have to work in paid employment

- Earn wages which are primary to the household

- Occupy working classes

Gender divisions in paid work

Women:
- Perform 'routine' administrative reception, sales and clerical work which is poorly paid; or
- Work in production in the noisy weaving and winding rooms 'minding' machines; or
- Completed 'unskilled' mending/ sewing work in the all-female mending room

Men:
- Held managerial jobs earning high salaried working in plush, quiet offices; or
- Completed middle management in production, office, sales and design work; or
- Worked in the dye house and wet finishing areas involving dirty, dangerous, heavy, hot and steamy work; or
- Worked in 'technical' tasks in the noisy weaving and winding rooms (while the women minded the machines);
- Dominated stores and dispatch areas

(Continued)

Box 4.1 (Continued)

Sources: Johnson, 1991 and 2000: 70–2.

These geographies have been important for highlighting the gendered experience of work and associated social responsibilities that differ for men and women. Moreover, geographers working in different settings have noted the different spaces and socio-economic environments involved in these gender divisions and gender relations that shape men's and women's unequal experiences. The notion of gendered spatial relations has therefore emerged as a second key theme in the socio-material geographies of gender.

GENDERED SPACE

As geography was affected by both spatial science and critical approaches, early socio-material geographies of gender also documented the specific gendering of space, especially urban space (England, 1991; Little et al., 1988; McDowell, 1983; Mackenzie, 1988). Cities were acknowledged to be structured in ways that affected and maintained women's secondary, ancillary, reproductive and service positions in society (Bruegel, 1973; Burnett, 1973). Home and suburban spaces were seen to confine women and separate them from opportunities in productive and public spaces of the city, including through housing markets and policies and landuse planning (England, 1991; Hayford, 1974; Little et al., 1988; McDowell, 1983). The impact of social changes and gendered negotiations in work and households is nevertheless creating change in urban spaces. Thus spatial arrangements of residential, employment areas and transport networks were shown to reflect traditionally uneven gender relations, but are also seen as dynamic and changing. For instance, England argued:

[T]he housing industry is still geared toward the 'traditional' nuclear family and has yet to adequately respond to the massive increase in women householders and the feminization of poverty by providing affordable, liveable housing. As a result, the patriarchal nature of the spatial structure of the city is permanent, as least for a while. ... [However] women are actively changing the spatial structure of the city. Since the 1960s women have been pushing for changes in land-use and zoning legislation, alternative housing structures, and neighborhoods to better fit their new roles and responsibilities. (England, 1991: 143)

In a complementary way, rural geographers have highlighted how village and farm spaces have also been gendered – usually in conservative or patriarchal ways. Stebbing (1984), Little (1987, 1994b) and Whatmore (1991) have shown how women are expected to confine the majority of their activities to domestic or nurturing spaces (e.g. barns for the care of young or sick animals) and to the sites where rural communities are sustained (e.g. village shops, fund-raising sites, community halls and the kitchens and service spaces of sports facilities).

The documentation of gender and space in terms of urban and rural environments has predominantly focused on the different ways men's and women's day-to-day lives and opportunities intersect with the expectations and meanings that are attached to different settings. There is, however, an additional set of work that has drawn on both radical feminism and previous environmental studies to consider the safety and gendered contestation of space. This literature has considered women's safety in different spaces and the way spaces carry social meanings or threats that influence women's behaviour and experience. Drawing on radical feminist sociologists' and criminologists' attention to the violent expressions of patriarchy, geographies of crime and fear began to document the spatial patterns of women's experiences. S.J. Smith's (1987) early work on this topic showed that fear of crime affects when and how women move about cities. She also showed how women act to reduce their fear of crime, confining their activities at night and using their homes as a refuge from crime. Experience of crime was not the only factor in shaping the way women negotiate space. Indeed, more recent studies have shown more subtle associations between spaces and social meanings and behaviour, including fear of crime and social meanings constructing public streets and spaces as potentially threatening and women as potentially vulnerable (see Box 4.2). For instance, in Adelaide, Australia, Hay (1995) concluded that women were twice as unwilling as men to go to a shop, public phone, friend's house or public leisure site at night. Similarly, in Britain, Valentine's work (1989, 1990, 1992) has documented how public spaces can be scripted as inherently dangerous or threatening for women (a process by which aggressive and threatening gender relations are normalized into so-called 'public' spaces). Valentine also highlights how the positioning of crime and fear in the public sphere reduces the focus on homes as sites of violence and domestic crimes against women.

4.3.3 Socio-cultural geographies of gender: identities, performances and struggle over femininity and masculinity

The material conditions and expressions of gender relations continue to shape the everyday lives of men and women. Since the 1990s, however, the increased interest in cultural theory, poststructural and postcolonial approaches to geography have meant that scholars have focused more on gender as socially and culturally constructed sets of meanings and practices. Initially these meanings developed from an appreciation of the heterogeneity and differences that divide gendered experiences across 'class or racial or other variables' (Pratt and Hanson, 1994: 12), but attention to meanings of masculinity also developed quickly throughout the 1990s. Thus, while the physical realities of work, access to housing, services and negotiations of space continue in the lives of men and women in all societies, social geographers have increased their focus on the discursive and cultural politics that ensure gender involves a diversity of meanings about maleness and femaleness. This work has highlighted the way meanings and practices are assembled, produced and circulated to convey dominant gender identities. And as this section will show,

---

Box 4.2 Gender, space and fear: 'dangerous places'

Across many studies, women have noted their avoidance or fear of many activities that involve negotiating various public spaces – especially at night. In some cases, these spaces are the environments that must be traversed for day-to-day activities; in other cases they involve specific public sites that are supposedly facilities available for general use of all people. Finally, there are key environments that are perceived as particularly dangerous spaces as a result of their physical (and socially isolated or threatening) character.

Activities and spaces that were associated with fear include:

- Walking as a mode of moving to/from a day-to-day location
- Walking to the shop
- Walking to a friend's house
- Walking from a cinema, bar, restaurant

- Using 'public' facilities
- Using public phones
- Using multi-storey car parks
- Using public transport

- Dangerous settings
- Traversing closed-in pathways
- Traversing deserted spaces
- Traversing parks or other open spaces

While considerable work has been done on studying the physical attributes of these spaces and planning to manage safety in different spaces, geographers have also been instrumental in highlighting that it is the social relations and gendered assumptions that shape people's experiences of spaces as safe or dangerous, and not just the physical character of the spaces.

Sources: Hay, 1995; Pain, 1991; Valentine, 1989, 1990, 1992.

---

geographers have investigated the variety of social meanings, cultural processes and physical and imagined spaces that are involved in construct-ing (or challenging) dominant notions of femininity and masculinity. Further discussion and examples of these trends are provided in Part V that outlines the changes Jo Little has made in her work on gender and geography.

CONSTRUCTED AND NEGOTIATED FEMININITIES:
DIVERSITY, IDENTITY AND ALTERNATIVE POSSIBILITIES

Geographers moved into the 1990s recognizing that gender involved both a form of uneven social relations and a range of processes that produce **gender identities**. Whatmore explained: 'Rather than simply elaborating or expressing 'natural' differences, social processes, through a range of institutions, codes and taboos, mediate and modify biological differences and inscribe them in the gender identities of masculinity and femininity' (1991: 34). Consequently, feminist geographers have worked to document the ways institutions and industries have intersected with cultural values and beliefs to construct forms of femininity (and masculinity). Examples of this type of work are summarized in Table 4.2. Initially, drawing on earlier foci on paid work, Pratt and Hanson (1994) were able to show how work- and place-based differences contributed to the formation of gender identities across four different communities in Worcester, Massachusetts. They concluded that '[d]ifferent gendered identities congeal around women living in different areas'. One of their comparisons illustrates this contention. In the 'impoverished' inner neighbourhood of Main South where many Spanish-speaking and Asian immigrants worked in manual factory work, Pratt and Hanson discovered women were valued for their stability and tolerance of boring, repetitive work requiring manual dexterity. In contrast, in the relatively affluent, white, middle-class area of Westborough, women were defined as hard-working, dependable and loyal in positive and respect- ful ways that Pratt and Hanson observed as 'wifely' characteristics.

Drawing on Pratt and Hanson's work, Little (1997) shows a different way in which gender identities are formed in rural English villages. While recognizing the diversity of women in two villages (in terms of length of residence, class, ethnicity, age and disability), Little argues that feminine identities are ordered (affirmed or discouraged) according to the degree to which they support dominant or idyllic ideas about country life and rural communities. Dominant identities of *where* rural women should be, and what work they should do, results in their strong association with family, home and community, and limits their participation in paid employment. While in specific villages these locations for women include sites such as shops and village halls, they also involve the imagined space of 'home' and 'community'. The association of gender identities attached to these spaces acts to restrain women's choices about paid work. She concludes:

> [T]here are a number of characteristics that have assumed a powerful and central role in the expectation and assumptions of women's iden- tity in rural communities ... a set of powerful assumptions which influence beliefs about gender identities in rural communities. ... Internal and external beliefs about the appropriateness of women's involvement in childcare and family reproduction has been priori- tized over employment. [Also,] the idea of women as 'community- makers' has ... assumed a central role in the identity of rural women. This idea has been reflected in a number of ways including the participation of women in voluntary work and the organization by women of village events – again affecting their involvement in paid work. (Little, 1997: 155)

TABLE 4.2   FEMININITIES: CONTRASTING FORMATIONS OF GENDER IDENTITIES

Processes of gender identity formation	Empirical setting	Geographers' writings
Women's identities are associated with different types of work and neighbourhood location within industrial and middle-class areas of a city.	Four urban residential and work sites: Worcester, USA	Pratt and Hanson (1994)
Women's identities are constructed in ways which strongly link them with idyllic rural meanings shaping 'home', 'community' and 'employment' in English rural society.	Two rural villages: Avon, England	Little (1997) – see also Little (1987) and Little and Austin (1996)
Women's identities are read and performed through certain expectations about 'bodies' (type, shape, dress, behaviour) associated with certain street 'spaces' in a New Zealand city.	A main capital city street: Wellington, New Zealand	Longhurst (2000a)
Women's and men's identities are formed through dominant discursive accounts of 'farm' and 'industry politics' that are circulated through both rural media and farming organizations in Australia and New Zealand.	Media and farmer organizations: Australia and New Zealand	Liepins (1998b)

While work (for families or in paid employment) has continued to be a key focus of many gendered geographies, further poststructural accounts of gender identities have shown how different discourses and spaces are entwined in forming notions of femininity. This work has shown that identities may be fluid or changeable, and can be negotiated in a variety of transitory or ongoing cultural processes to form a politics of gender. Cameron (1998) illustrates how this can happen as Australian women negotiate work and subject positions within their homes and the prevailing discourses of domesticity and femininity (see also Hughes, 1997 for a rural example).

In two rather different examples, summarized in Table 4.2, Longhurst and Liepins document how women are associated with both embodied and discursive identities that frequently associate certain forms of femininity with particular bodies, places and actions. Longhurst has developed a major interest in the body as a site and space through which gender is performed (Longhurst, 1995, 2000a, 2000b, 2001; Longhurst and Johnston, 1998). Taking the case of bikini-wearing pregnant women participating in a beauty pageant, Longhurst shows that:

- Ideas of femininity include meanings and expectations about *bodies* (where bodies become corporal spaces through which feminine identities are assembled).
- Identities are *read* and expected to operate in prescribed ways in different spaces (e.g. pregnant women acting modestly and demurely in public spaces).
- Identities can be *performed* in ways that challenge spatial and social expectations (e.g. bikini procession across a main city street destabilizing norms about pregnant femininity – see Figure 4.1).

FIGURE 4.1   BIKINI BABES – MIXING AND PERFORMING GENDER IDENTITIES

Source: '"Bikini Babes" – Contestants in the pregnant bikini contest … halt traffic on Lambton Quay this morning'. Photograph by Phil Reid in *The Evening Post*, 7 October 1998, page 1. Reproduced by courtesy of The Dominion Post (see Longhurst, 2000a for a discussion of this contest and the performances of gender involved).

In contrast, I have documented the way women involved in farming have their identities constructed in predominantly conservative ways by both farmer organizations and rural media (writing as Liepins, 1998a). This is an example of how cultural processes weave identities and spaces together to form social expectations and reiterate assumptions about gender. Despite their considerable contribution to agriculture, women's identities are rarely associated with agricultural industry and politics but are most frequently associated with relationships and farm spaces involving care and support, whether this is of animals, children or male farming partners (though some exceptions, such as feminine gender identities linked to cheese-making, exist – see Morris and Evans, 2001). In other work I have documented the ways women's identities as farmers have been negotiated and reconstructed by farm women's groups in Australia (Liepins, 1996, 1999; Panelli, 2002a). These groups actively seek to increase the visibility and involvement of women farmers and have strategically constructed textual and visual expressions of that identity.

Overall, these newer studies of gender identities provide additional avenues for us to understand how meanings and expectations of femininity are constructed and contested. These developments are further explored in Chapter 7.

'MAKING MEN': IDENTITIES BOTH DOMINANT AND ALTERNATIVE

Complementing the gender identity work focused on women, a number of geographers have also begun to interrogate the social constructions and cultural processes that support particular understandings of masculinity. This has been an important development in the 1990s, complementing the earlier feminist geographies, for it has turned a critical gaze on to the dominant gender in most social settings, asking what processes produce and maintain men in powerful positions in terms of gender relations. With the rise in interest in poststructural and cultural theory, geographers have concentrated their efforts on how constructions (and rebuttals) of dominant masculine identities are assembled and circulated (see Figure 4.2). This section reviews some of this work and closes with a discussion of recent studies showing the complexity of masculine identities even while some remain particularly dominant.

A focus on the cultural processes by which masculine gender identities are established resulted in early work analysing the textual constructions of masculinity within different media arenas. Jackson's (1991, 1994, 1999) work in the British context showed that textual and visual representations of masculinity were powerful ways in which some identities and meanings were produced and maintained as dominant, even while the possibility exists for other masculinities. Jackson notes that:

Hegemonic masculinity asserts the 'naturalness' of male domination, based on solidarities between men as well as on the subordination of women. It is rooted in essentialist notions about inherent biological

(a)

(b)

FIGURE 4.2 CONSTRUCTING MASCULINITY THROUGH BEER ADVERTISING: THE 'SOUTHERN MAN'

Note: The 'Southern Man' is a commercial version of a particular form of rural masculinity attached to ideas of a rugged, outdoor, agricultural lifestyle in a specific region and landscape of New Zealand: gender and place are combined in a widely known icon of the Speights beer commercials which occur in both urban and rural settings (Law, 1997; see also http://www.speights.co.nz/south_southernman.cfm for the 'Southern Man Identification Chart').

differences between men and women, from which all kinds of social consequences are alleged to follow. ... In privileging certain socially-approved forms of masculinity, other forms are implicitly subordinated. By their insistence on an exclusive heterosexuality, for example, dominant masculinities impose a variety of sanctions on homosexual and bisexual practices. ... Dominant masculinities are [also] oppressive of heterosexual men who may not wish to live up to masculinist ideals of emotional self-control, intellectual rationality and sexual performance. (Jackson, 1991: 201)

Jackson's own studies of social texts (e.g. books, movies, posters, magazines) have been extended by other geographers' attention to a variety of cultural processes and discourses. For instance, Berg (1994) and Nairn (1999) have shown how hegemonic ideas of masculinity have affected how geography has been understood and practised as a field science. In a different fashion, analyses of rural print media have illustrated how men are 'made' as rugged farmers and powerful industry leaders (Liepins, 2000b), while Jones' (1999) and McDowell's (2002a) respective attention to literature and policy discourse has shown how ideas of masculinity are also constructed around childhood and young adulthood. Different aspects of these works are summarized in Table 4.3 but each case illustrates how popular and professional discourses associate masculinity with meanings such as strength (physical force through to mental prowess) and control (formal authority through to subversive laddishness). Other recent work has also noted the association of hegemonic masculine identities with different national imaginings (Morin et al., 2001; Tervo, 2001, as discussed in Chapter 7).

While the effect of cultural theory has been widespread in social geography, attention has not been exclusively textual or symbolic – removed from the everyday lives of men (and women). Indeed, the themes of work and space, noted in the earlier geographies of gender, continue to be relevant and useful to these more cultural/identity-focused studies. For instance, continuing the feminist interest in the gendering of public space, Day (2001) has extended the literature of geographies of fear and safety by showing that gender identities are closely involved in men's ideas about women in public space. In her study of Irvine, California, she showed that despite Irvine being considered a relatively safe place, young male students held perceptions of women as vulnerable or endangered in public space. She argues that this is in part an expression of two masculine identities, one being the 'youthful "badass"' and the other being a 'chivalrous man'. This example illustrates the way identities are assembled and repeated in our understandings of how gender and space intersect.

Other space-sensitive geographies of masculine identities include analyses of exclusively male spaces. Smith and Winchester (1998) highlight how men's groups can provide social spaces through which men can explore and negotiate identities and the possibilities beyond the constraints of

TABLE 4.3   CONSTRUCTING MASCULINITY: PROCESSES AND MEANINGS

Constructions of masculinity (meanings)	Processes by which masculinity is constructed	Geographers' work
Outdoor, rugged, exploring, (male) geographers conduct geography as a 'field-science'.	Constructed meanings are established through the practice of fieldwork and 'exploration' in geography.	Berg (1994) Nairn (1999)
Outdoor, rugged, strong, controlling farmers, are the central figures in agriculture.	Constructed meanings are assembled and circulated through the photography and texts produced in rural media.	Liepins (2000b)
Working class young men are understood as 'loutish', 'yobs' and 'lads'.	Constructed meanings are articulated through popular and professional discourses found in the media, as well as government discourse on education and employment.	McDowell (2002a, 2002b)
Young boys are identified as adventurous 'natural' figures in stories of country childhood.	Constructed meanings are created through imaginary subjects in popular cultural portrayals of childhood in British country areas.	Jones (1999)

hegemonic masculinity that Jackson defined. Taking the example of men's groups in Newcastle, Australia, Smith and Winchester have shown that while many of the middle-class men valued their work achievements and participated in articulating many norms of hegemonic masculinity, they also sought alternative ways of negotiating gender at home with children or partners. Their groups provided a place to express the pressure, conflict and stress of negotiating different wishes and ways of being male.

> The role of the men's groups was ... to provide a negotiating space, a place where men could be listened to, where they could be supported, and a place where they could discuss 'over and over and over again' the negotiation of gendered place-based identities. The men in the men's groups made for themselves an informed position of negotiation. (Smith and Winchester, 1998: 338)

This work overlaps with a number of studies that continue to use the lens of work as a means to view and understand the construction of, and struggle over, gender. Work by Massey (1996) and McDowell (2002a) illustrate how different workplaces involve the negotiation and performance of certain gender identities. These works are interesting for the different industries and class contrasts they provide. In the case of 'high-tech' industry, Massey recorded how masculinity was constructed in association with this work in terms of the long hours, use of high-powered computers, use of reason, 'scientificity' and abstract thought. In contrast, McDowell's study of young working-class males showed that masculine identities were assembled in association with work (rather than schooling) and with particular industries (manufacturing rather than retail), as well as an expectation of responsibility for future dependents:

> [Y]oung men in both cities had clear ideas about the nature of work that they regarded as appropriate for men, despite the changing realities of the local labour markets. Their strong views about the gendered characteristics of many service sector occupations led them to reject some occupations as 'girl's work'. ... [T]he majority of young men ... expressed preferences for manufacturing employment. ... For these young men, finding and holding onto waged work was a serious matter. ... [They] exhibited a more serious sense of themselves as dependable workers, seeing themselves at the start of a long period of respectability and masculine responsibility for dependents. (McDowell, 2002a: 110–11)

Overall, these recent studies of masculinity have been fruitful for extending geographies of gender in two ways. Not only are notions of identity increasingly employed (as further discussed in Chapter 7), but also understandings of gender are being widened as traditional attention to women is contextualized in terms of the way dominant (and alternative) notions of masculinity are constructed and articulated through popular discourse, space and work.

4.4 CONNECTIONS AND CHALLENGES FOR GEOGRAPHIES OF GENDER

Geographies of gender commenced with an explicit focus on women – both within and beyond the discipline. But just as Chapter 2 showed that social geographies have been influenced by a range of different approaches over the last 30 years, so too the study of gender has been shaped and invigorated by several major currents in theoretical thought and politics. One account of how these dynamics are traced through the work of an individual geographer is produced in Part V where Jo Little's work has moved from 'adding women' to her research to a point where she has investigated the power relations and the processes by which both material and (more lately) cultural expectations affect women and men and our notions of femininity and masculinity. In a more general way, this chapter has shown how an explicitly feminist agenda has been important for showing the material and cultural ways societies are gendered. Early work focused on the uneven arrangements and values placed on men's and women's work, service needs and access to space. But more recently gender geographies have drawn – and faced challenges from – other research and theory. These developments form six connections and/or challenges:

First, critiques of gender in the practice of geography continue both within and beyond social geography. However, even those works that do not explicitly identify as social geographies work as important resources for social geographers who wish to continue to critique their practice and the strategies and choices they make in conducting their work (McLafferty, 2002).

Second, intersections between economic and social geographies (with both urban and rural emphases) continue to provide work-based questions that geographers are pursuing as a way to understand the continuing material inequality and power relations that differentiate many men's and women's lives (Blumen, 2002; Morris and Evans, 2001; Tonkin, 2000).

Third, acknowledgement of difference through postmodern and postcolonial studies continue to challenge gender analysts to define the categories of gender that are used and the diversity and inequality that fracture broad notions of 'women' or 'men'. In different ways, geographers have pursued intersections between gender and class, ethnicity and/or sexuality, but far more can yet be harvested from these projects (Blumen, 2002; Pratt, 2002).

Fourth, connections between social and cultural geographies of gender are providing powerful new ways of understanding how certain gender meanings and identities are privileged and dominate available repertoires for individuals and groups to consider – be they workers, magazine readers, school leavers, etc. (Little, 2002a; McDowell, 2002a, 2002b).

Fifth, drawing on geographies of sexuality and identity, recent studies of gender have extended to include greater attention to the embodiment and bodily negotiations of gender (Johnston, 2002; Longhurst, 2001).

Finally, a continued connection between the social critique and politics of gender sees feminist geographers' ongoing interest in documenting the position of women within geography (Hanson, 2000; Luzzadder-Beach and Macfarlane, 2000; Winkler, 2000) and the strategies and actions women will make to move beyond prescribed identities to the conditions and recognition they seek for themselves and others (be this within everyday life or the field of critical geography) (Cameron, 1998; Fincher and Panelli, 2001; Longhurst, 2002).

SUMMARY

- Gender has been a widely acknowledged category of social difference in geography since the 1980s (although some important initial pieces were also written in the 1970s).

- Reflecting the positivist/liberal, critical and poststructural tides that have influenced geography and feminist studies more widely, notions of gender have been conceptualized respectively in theories of gender roles, gender relations and gender identities.

- Gender is frequently recognized as the experiences and expectations people negotiate as a result of social and cultural meanings attached to maleness and femaleness.
- Ideas, assumptions and power relations surrounding gender do not only affect men's and women's experience socially, but also their economic and political interests, circumstances and opportunities.
- In geography, the first gender analyses stemmed from feminists recognizing the absence of women as subjects of research and the difficulties women faced within the discipline. Work focusing on 'women in geography' resulted. This included accounts of women geographers operating within different national academic settings. It also included calls to add women as research subjects into existing human geography research programmes more explicitly.
- Geographies focusing on the socio-material experiences of gender followed. These works concentrated on gender divisions and relations that resulted in the unequal patterns of work, housing and experience of space in everyday life.
- More recently, socio-cultural dimensions of gender have been investigated to show how meanings and power relations surrounding gender are constructed and contested through both identities of femininity and masculinity. This work has coincided with the rise in studies of masculinity generally, and with postmodern attention to diversity, and poststructural attention to the discursive construction and negotiation of gender.
- Intersections with Marxist and class-based analysis continue to be important for understanding the unequal material and socio-economic conditions shaping women's and men's lives.
- Use of postmodern, poststructural and queer theory enable geographies of gender to more fully engage with the heterogeneity of gender categories and experiences, such that intersections with class, ethnicity and sexuality can be pursued to appreciate the differences between women and between men at bodily, material and cultural levels.
- Further developments in geographies of gender illustrate ongoing questions and fields of inquiry via continued attention to economic inequalities and gender politics as well as new foci on bodies, sexuality and performativity.

Suggested reading

One useful way to approach gender is through an understanding of different feminist theories – Tong's (1992) book provides a clear and detailed account of different liberal, socialist, radical and postmodern perspectives while some geography textbooks give useful summaries (see Blunt and Wills, 2000: Chapter 3; and Waitt et al., 2000: 83–92). A different approach can be taken by working outwards from recent theories of masculinity – Connell's (1995) and Whitehead's (2002) books both provide useful starting places – while geographers have tended to concentrate on specific studies of masculinity in particular settings (highly readable examples are given in journal articles such as Jackson's (1991, 1999) investigations of how masculinity is constructed in discursive and economic processes such as advertising and male lifestyle magazines; McDowell's (1995 and 2002b) analyses of masculinity as it is performed through work; and Woodward's (2000) account of masculinity in military training). Finally, studies of gender are regularly updated in progress reports in the journal *Progress in Human Geography* (see, for example, Longhurst, 2002) and a wide range of theoretical and empirical studies are regularly published through the journal *Gender, Place and Culture*.

5 Race and Ethnicity

5.1 INTRODUCTION

Recently a group of my third-year social geography students designed and presented a seminar on 'Geographies of Ethnicity'. In discussing their plans with me, one said:

> We want to show different approaches? Like, not just the different topics studied, but *how* it was done.

I agreed that the 'how' question – or the practice of geography was important. So during the seminar the presenters summed this up potently, and in ways their peers related to closely. After discussing a piece on research with Maori (McClean et al., 1997), the closing speaker said:

> Fifty years ago, I could've studied you, by your race and your phenotype. And you could've studied me – [using] Maori statistics: lack of education, unemployment, poverty. ... But these days you gotta do things differently. You gotta recognize things [knowledge and values] that might be different [to what you know]. It's like you bring lots of Western theories into it. ... That's OK. But you gotta *see* that you're doin' it, eh. ... You have to think about other things like **whānau** – family's important. [You can't just] count people or measure or interview them by themselves – 'cos they live in a context [as part of a whole *whanau* group]. ... And *aroha*, lots of people just take that as 'love', but for research it means doin' it with care, respect. Caring about why yer doin' it, like being passionate. Haaah [laughter] well I mightn't go that far. But caring about it – and who yer talking to and why yer writing that geography. (Excerpt from class presentation, 2000)

This forthright and down-to-earth conclusion was a well-placed closing statement from the presenters. These students had distinguished between **race** and **ethnicity** in their session but they also knew geographies of ethnicity had changed to become more reflective about what knowledge and research assumptions were being brought to bear in such research. And they closed with an example of how the significant values of different ethnic groups (such as Maori) could contribute to the practice of social geographic research.

This cameo provides an effective frame for the following chapter. First, note is made of how the key terms of race and ethnicity have been understood. Then, geographies of race and ethnicity are presented in order to highlight how different approaches to social geography have produced quite different products and enhanced our understandings of various dimensions of life as ethnically influenced. Finally, the chapter concludes by outlining some of the contemporary challenges and connections being made by geographers focusing on ethnicity and the way that it intersects with other social differences or challenges the practice of creating geographies.

5.2 DEFINING RACE AND ETHNICITY

Geographies of race and ethnicity have an extensive history as scholars have long documented and classified different populations according to their racial origins, ethnic cultural differences and the territories and places that are significant to them. Nevertheless, a heightened sensitivity to questions of race and ethnicity has developed as quantitative, Marxist and feminist geographies have been criticized for their Anglo-western assumptions and homogenizing constructions of knowledge. Careful consideration of the two terms is therefore a necessary starting point.

5.2.1 Race

Definitions of race originate from historical classifications of people according to phenotypes (categories of physical difference) that were linked to different genetic makeup or 'blood'. Racial classifications were closely linked to social Darwinism and scientific racism that developed through the last 200 years (Miles, 1989; Spoonley, 1994). In this environment, people were labelled by race but then the racial classifications were linked to other supposedly related qualities such as degree of civilization, industriousness and mental ability (see Box 5.1). Anderson explains:

> Phenotype was gradually added to the European cognitive package, and by the late seventeenth century, skin colour was an independent justification for the enslavement of American 'Negroes' ... [and by] the mid-nineteenth century ... Western scientists [were] arguing that a *naturally* rooted relationship existed between phenotype, culture, and civilizing capacity. Not only that but a chain of 'races' existed, at the apex of which stood the white race as the supreme measure of biological and cultural worth. (Anderson, 1988: 129)

Box 5.1 An example of racial classifications

Many geography schoolbooks from the early twentieth century illustrate how race was used to classify people. Duce's text on Africa was first published in 1935 but was used extensively throughout the British Empire/Commonwealth and was reprinted in 1946, 1951 and twice in 1956.

This text included descriptions of different people, classified in a ranking of races. Judgements were made about different types of people according to the regions and climates in which they lived. These descriptions are associated with a geographical approach known as environmental determinism. Examples include a contrast between tropical and temperate areas.

TROPICAL LANDS

Here the lands are very hot; the fall of rain is very heavy and occurs at all seasons of the year. ... [It] is very difficult to find people who can and will work hard in such humid conditions. (1956: 50)

TEMPERATE OCEANIC LANDS AND FORESTS

Here are the parts of continents in temperate lands which receive warm, moist, westerly winds all the year. ... These lands are principally the homes of white peoples; and since the summers and winters are mild work is possible all the year round. The need for providing for the winter has created an active, vigorous people, who are the most forward in the world. (1956: 55–6)

Addressing African populations specifically, Duce employed racial classifications even more strongly, to differentiate between 'better' and 'poorer' races. For example:

It is in dense forest country only that the backward negroes live, driven there by the more active natives. Among these backward people are the Pygmies. In every way, mentally and physically, they are a poor people. ... The better types of natives are called Bantus. ... They are typical negros to look at, leading a simple savage life. ... They show much skill in carving in wood and ivory; and pottery-making and simple weaving are common to them. (1956: 75, 76, 78)

'Pygmies and a Normal Native'

Racial classifications became problematic for a number of reasons, as shown in Box 5.2. Official uses of the term 'race' have declined in science and international relations as a result of these challenges. However, common discourse and discrimination based on 'race' continued because of the powerful sets of knowledge that have been constructed from *racialized* categories. Racialized discourse includes assumptions about racial purity, moral worth, mental capacity or social status, even though the scientific validity of such knowledge has been successfully challenged. And as section 5.3.3 will show, racialized groups must negotiate the implications and prejudices of these discourses when they seek to organize basic living arrangements such as employment and housing (Gilbert, 1998; Rees et al., 1995; Zavella, 2000).

Box 5.2 Race becomes problematic through the twentieth century

IMPLICATIONS OF RACE IDENTIFICATION
- Genetic and physical attributes were taken as indicative of individuals' capacities or qualities (e.g. intellectual ability, social status or moral worth).
- Entitlements or constraints (e.g. to citizenship) were meted out through documentation of racial 'purity' ('full blood', 'half-caste').
- Racialized classifications were constructed and apportioned for ideological and political reasons (e.g. American slavery and genocide of Jews and Gypsies).

PROBLEMS CHALLENGING THE USE OF RACE IDENTIFICATION
- Racial boundaries are increasingly challenged by genetic mixing.
- Racial 'correlations' between physical differences and other attributes (e.g. intellect) are shown to be scientifically flawed.
- Racial definitions are increasingly rejected as politically divisive.
- Racial definitions are increasingly registered as a basis for discrimination or the denial of human rights.

An example of racialized discourse can be seen in Maddrell's (1998) study of how geographical knowledge has been constructed and disseminated in schools. Taking the case of historic geography schoolbooks (1830–1918), she shows how texts not only reflect but perpetuate relations of power that are racialized. Euro-centric comparisons were constantly made in these school texts such that all else was judged against European and English norms, including physical appearance, beauty, degree of civilization, urban structure, street design, and the social and political order and status existing in different societies. Moreover, Maddrell shows that these texts were integral to notions of the 'right to rule' – a belief that was central to the rise of colonialism. She explains:

School geography texts played a significant role in making known and legitimating the Empire to working-class children in Britain. ... In

Stewart's *A Compendium of Modern Geography* (1850, 9th edition) ... the text eulogises the industry and social practices of England and presents British dominance overseas as the inevitable product of domestic industry and export trade. (Maddrell, 1998: 91)

These racist geographies continued as works such as Duce (1935/1956) were widely used through the 1930s–60s to teach school children racialized facts (as was seen in Box 5.1). Today, while scientific and geographic identification of people according to 'race' is widely accepted as unhelpful and discriminatory, the term remains common as a way to identify people according to physical differences. Moreover, racially-based assumptions continue and a considerable geographic literature critiques the occurrence of different forms of racialization and cultural and institutional racism in urban and rural settings (Agyeman and Spooner, 1997; Anderson, 1988; Berg and Kearns, 1996; Murdoch and Marsden, 1994; Spoonley, 1994; Zavella, 2000). These investigations show how race is constructed through institutions, discourse and material conditions so that some groups are disadvantaged in settings such as legal and education systems, housing and labour markets, and census statistics. The racialization of groups and the creation of racism are shown to be social and spatial processes that geographers can critique. Examples of these works are discussed in section 5.3.3.

5.2.2 Ethnicity

Since problems surround the concept of race, ethnicity has been widely adopted as an alternative term and analytical concept. Spoonley explains:

> The idea of ethnicity, or a consciousness of cultural identity, provided a positive and less politically suspect way of classifying people into groups. In the case of ethnicity, group affiliation was self-claimed and not imposed according to phenotype as occurred with 'race'. (Spoonley, 1994: 85)

The term ethnicity therefore allowed both a degree of self-selection and a consideration of diverse socio-cultural characteristics supporting people's ways of life. Rather than classifying and deducing attributes of people based on a single criterion (of physiological difference) ethnicity allowed the acknowledgement of 'both a way in which individuals define their personal identity and a type of social stratification that emerges when people form groups based on their real or perceived origins' (Johnston et al., 2000: 235).

These perceptions are mutually constituted through ancestry, culture, traditions, beliefs and behaviour. Consequently, geographers working on ethnicity have acknowledged processes of racialization but have concentrated on the relations and actions shaping ethnic affiliation as it develops from inside a group. Ethnicity is therefore recognized to influence processes such as migration, residential patterns and identity politics

(Pulvirenti, 2000; Robinson, 1993; Valins, 1999). Examples of these works are discussed in sections 5.3.2 and 5.4.

5.3 CONTRASTING GEOGRAPHIES OF RACE AND ETHNICITY

As has been explained in Chapter 2 (and illustrated in Chapters 3 and 4), social geographies have been written from many contrasting philosophical and theoretical positions. Geographies of race and ethnicity are no different in this respect. For instance, the modernist and applied nature of quantitative geographies have been important for modelling and analysing patterns of racially- or ethnically-defined minority populations. In contrast, humanistic, feminist and poststructural works have produced geographies showing how different groups experience and make sense of the spaces and racialized social systems and discourses they encounter. A selection of these different approaches is outlined in this next section. An illustration of how the different approaches can be layered within one geographer's work over time is also provided in the example of Kay Anderson's work – see Part V. This example documents her move away from geographies that had mapped ethnic ghettos through to a more poststructural reading of the discursive and institutional ways racial meanings and inequalities become embedded in societies and places.

5.3.1 Mapping ghettos and socio-economic patterns

The development of post-Second World War geography as a **spatial science** and the rise in urban geography led to many quantitative geographies of urban segregation and ghettos. While the notion and analysis of ghettos has waxed and waned in the following 60 years, geographers' interests in mapping and analysing spatial and socio-economic patterns of racialized or ethnic minority groups has remained common.

Influential work in the 1960s focused on documenting the location and extent of ghettos. In 1965, Morrill published a Seattle-based study of 'The Negro Ghetto' that combined mathematical practices of the day with a focus on a specific social 'problem'. In reflecting on this work, Morrill (1993: 352) says: 'I was an ardent advocate of the "new scientific geography"'. Opening the 1965 paper, Morrill stated: 'The purpose here is to trace the origin of the ghetto and the forces that perpetuate it and to evaluate proposals for controlling it' (1965: 339) – see Figure 5.1. While Morrill (1993) later stated his strong social feelings about racial discrimination and residential segregation, the cultural demands of constructing geographic science in the 1960s meant that he confined himself to the mathematical techniques promoted within his academic environment. As shown in Figure 5.1, Morrill limited his work to mapping, describing and predicting spatial concentrations. Later he proceeded to review the process of ghetto

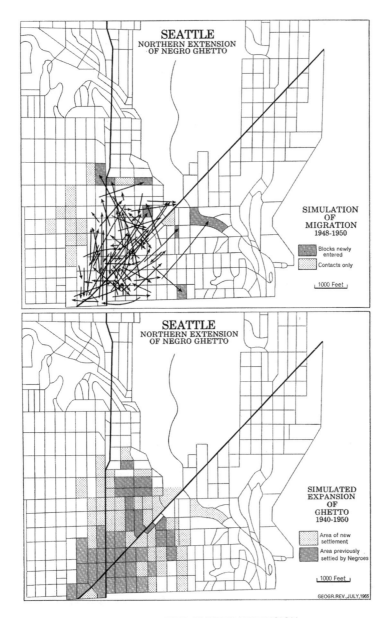

FIGURE 5.1   MAPPING AND SIMULATING GHETTO EXPANSION

Source: Morrill (1965: Figures 11 and 12).

expansion and proposed a spatial diffusion model to simulate how a ghetto would change/move/expand over time (Morrill, 1968).

The result of this modelling enabled some evaluation of Morrill's ideas and further general predictions were presented. While the technical and epistemological limits of Morrill's work were later criticized (see Jackson and Smith, 1984; Peach, 1993), the effectiveness of the modelling techniques and their relevance to topical social issues was described as 'classic' and as emphatically successful (Peach 1993; Rose, H.M., 1993). A measure of this success was reflected in its stimulation of other studies of ghettos and racial segregation (e.g. Berry, 1971; Robinson, 1981; Woods, 1970).

Mapping ethnic concentrations was a widespread practice in the latter part of the twentieth century. Even humanistic approaches to ethnicity, which concentrated on the need to understand groups' experiences and understandings through qualitative investigations, defined their field of inquiry through the mapping of different ethnic groups. For example, Cybriwsky's (1978) study of an inner-city area of Philadelphia in the early 1970s included the mapping and description of different ethnically-defined housing areas. This map was generated through participant observation, the analysis of surnames and the questioning of families about their ethnic affiliations. (As was common in the early 1970s, Cybriwsky's fieldwork was gendered in a way that privileged male experience – he recorded: 'In many cases, I asked residents to identify their ethnic affiliations and those of their neighbours. Maiden names of married females were not considered. Therefore, families were assigned the ethnicity of the husband' (1978: 20–21).) The resulting map showed how different groups of residents clustered together in blocks of relatively homogeneous ethnic affiliations, reinforced by different ethnic institutions and buildings (e.g. churches, schools, meeting rooms and businesses) that were maintained in the different locations. This type of mapping was to support later, more detailed, studies of how ethnicity and space were mutually constitutive as people's lives and actions were affected by (but also shaped) the spaces in which they lived.

Mapping and statistical analysis of ethnicity continue in contemporary social geographies, using various data sets from different national censuses. These geographies have predominantly described aggregate patterns of migration, residential location, housing status, household structure and employment in urban settings and although they have documented the often-difficult conditions of ethnic minorities, they have also highlighted the heterogeneity of different ethnic groups. For instance, Robinson's (1993) analysis of internal migration patterns within England and Wales shows contrasting patterns between West Indian, Pakistani and white populations. Similar contrasts are provided by Holloway (1998), who uses the USA census to show differences between black and white male youth in terms of their patterns of employment, continuing education or 'idleness'.

In complementary work, critiques of a black/white or white/non-white binary have been important in stimulating more nuanced studies of ethnic 'minority' populations (Zavella, 2000). In particular, Zavella has shown the highly constructed nature of ethnicity in the USA through census and citizenship policies. She has argued that the significant increase in Latino populations unsettles the black/white US identities that have dominated popular and policy arenas to date. Rees et al. (1995) provide a further example of non-binary, multi-ethnic investigations and the more detailed mapping and statistical analysis that can result from this type of work. Their investigation of 'Whites, Blacks, Indians, Pakistanis, Bangladeshis and Other-ethnicities' in Leeds and Bradford show just how different and complex socio-economic patterns can be. They trace demographic, household, employment, education, housing, and social class patterns and conclude that:

> We can no longer view the nonwhite population of Britain as a homogenous group ... [and a]lthough the nonwhite groups are concentrated in the same sorts of inner-city areas, examination of the maps and indices of dissimilarity show that they have quite distinct spatial patterns. ... To generalise, Indians have a class distribution with substantial and increasing middle-class elements; the Other-ethnicities have similar characteristics; the Black population has a more working-class distribution; the Pakistanis and Bangladeshi populations (and particularly the latter) find themselves in very difficult circumstances in the socioeconomic system with fewer qualifications, high unemployment, poor jobs and low incomes. (Rees et al., 1995: 589–90)

In a similarly comparative study, Byrne's (1998) analysis of Indians and Pakistanis shows that Pakistanis in Bradford were much more disadvantaged than both poorer and more affluent Indians in Leicester. But while the census data indicate these differences, Byrne must turn to wider factors, including 'ethnicity effects' to understand the patterns identified. Byrne notes that while locality factors (housing and labour markets) vary in the two cities, 'ethnicity effects' are particularly important because the Indians (mainly Punjabis or Gujeratis) and Pakistanis (mainly Mirpuris) have quite different access to capital and resources. This is further affected by different gender values that influence women's participation and income-earning ability in the labour-market. The increasing awareness of ethnicity-based social and cultural relations beyond the statistics is echoed by Fieldhouse and Gould (1998). They conclude a study of unemployment trends for white, black and Asian groups by stating:

> There is not a simple difference between white unemployment and ethnic minority unemployment, but ... for different ethnic groups there are different underlying patterns and processes which relate to the varied histories of migration, settlements, and employment; the different cultural and economic backgrounds; and the varying degrees and responses to discrimination and racism. (Fieldhouse and Gould, 1998: 850)

Not surprisingly, the limited explanatory ability of quantitative analyses of ethnicity have stimulated a wide range of other geographies that focus on the social and cultural relations within and between different groups. The following two sections illustrate two ways this has been pursued.

5.3.2 Social narratives and critiques of ethnicity and space

A contrasting literature complements positivist and statistical accounts of race and ethnicity by focusing on the lived experiences of different ethnic groups and their negotiation of space and racialized relations. Much of this type of work has stemmed from investigations of minority migrant groups whose ethnic identity is significantly different from the dominant resident population (e.g. African-Americans in the USA; Pakistanis in the United Kingdom; Italians in Australia). And, drawing on qualitative research traditions popular in humanistic and feminist approaches, many authors have documented individual's own accounts of life where ethnicity affects socio-spatial relations and/or experiences of racialization.

Gilbert's (1998) study of poor women's spatial boundedness in the USA illustrates how activities and spatial mobility of white and African-American women vary because of racism. She found that African-American women had more spatially confined activity patterns and their use of church and personal networks result in different (and more limited) access to work, housing and child care options (see Figure 5.2).

> African-American women's strategies in Worcester reflect the constraints they experience from individual and institutionalized racism in the housing and labor market. African-American women's use of rootedness – their reliance on networks, particularly of kin-based, and some African-American women's use of church in their strategies – supports the previous research that has documented the importance of the extended family and church in many African-American's lives as strategies for living in a racist society. (Gilbert, 1998: 614)

More recently, Zavella (2000) has shown how ethnicity shapes numerous spaces in the USA at contrasting scales. Tracing the emergence of Latinos as a distinct ethnic group in the USA, she documents how their identity has been racialized through different terminology. She then shows how Latino families and socio-cultural organizations are shaping household, regional and transnational spaces. The emergence of 'binational families' and 'binational political spaces' are two examples she uses to challenge the conventional (often binary-dominated) geographies of race and ethnicity, arguing: 'our thinking must go much beyond previous paradigms of race relations' (Zavella, 2000: 163).

Geographers' case studies have been important for countering the dominance of white/western/scientific views of the world and the spaces in which we live. They have shown how contrasting ethnicities will affect how people view their environments, construct knowledge and use space. In the case of New Zealand, Pawson (1992) reminds us of the contrasting values and cultural knowledge and human–environment relations different groups will hold. Comparing European and Maori views, he highlights

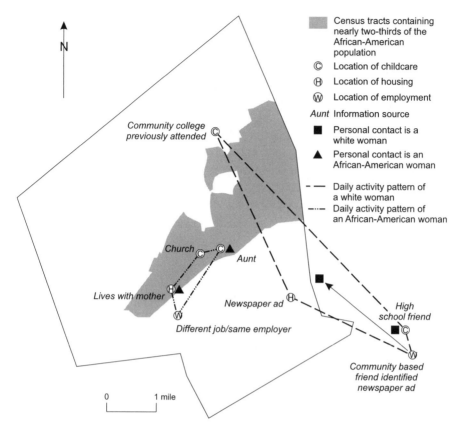

N

Census tracts containing
nearly two-thirds of the
African-American
population

© Location of childcare

Ⓗ Location of housing

Ⓦ Location of employment

Aunt Information source

■ Personal contact is a
white woman

▲ Personal contact is an
African-American woman

– – Daily activity pattern of
a white woman

–·– Daily activity pattern of
an African-American woman

Community college
previously attended

Church

Aunt

Lives with mother

Newspaper ad

High
school friend

Different job/same employer

Community based
friend identified
newspaper ad

0   1 mile

FIGURE 5.2 AFRICAN-AMERICAN AND WHITE WOMEN: CONTRASTING
NETWORKS AND SPATIAL RANGE

These two women's lives illustrate the contrast in African-American and white women's
survival strategies, the nature and spatial extent of their networks, and the spatial
extent of their daily activity patterns.

Source: Redrawn from Gilbert (1998: Figure 2).

the differences between hierarchical and competitive views of the former
and the interlinked and more holistic views of the latter (see Figure 5.3).
These affect both how ethnic groups view different people and also their
connection with the world (the physical environment and specific places)
around them. For example, Metge (1967) explains how the **marae** is a central
cultural and spiritual place in Maori culture, which has important generic
spatial structures and cultural values that contrast sharply with European
and Pakeha norms (see Box 5.3).

At an urban neighbourhood scale, Valins (1999) shows how ethnic identity can affect the cultural significance and practical arrangement of space.

99

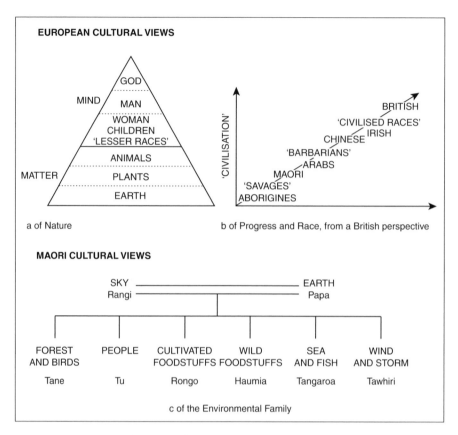

FIGURE 5.3 CONTRASTING VIEWS AND HUMAN-ENVIRONMENT RELATIONS

Source: Redrawn from Pawson (1992: Figure 2.1).

Taking the example of Broughton Park, an orthodox Jewish neighbourhood in Manchester, Valins documents how an urban residential area involves the interdependence of material space, imagined space and spatial practice at a variety of scales.

> Broughton Park is a place of connections. There are residents from across Europe, America and the Middle-East; telephone conversations take place with friends and relatives throughout the world; people come to study in yeshivot from across Britain, and as far as Russia and America; and marriages are even occasionally arranged via the internet. ... Through the construction of social and spatial boundaries, the practice of a highly regulated and ordered lifestyle, many orthodox residents actively seek to 'slow down' and institutionalise the spaces within which they live. ... Many orthodox

Box 5.3 The marae: an important cultural space for Maori

Joan Metge has written a number of important texts about Maori life, culture and spaces. This excerpt comes from one of her books. It is reproduced according to her own wording and views.

Today Maoris use the term *marae* in two related senses: first, for an open space reserved and used for Maori assembly, and secondly, for the combination of this open space with a set of communal buildings which normally includes a meeting house. Their meaning is usually clear from the context. For our purposes, it is useful to distinguish the former as 'the *marae* proper'. (1967: 173)

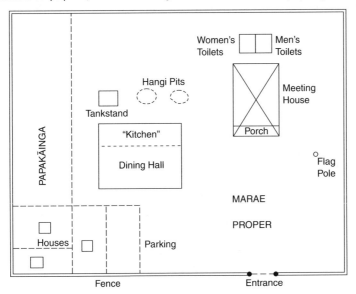

Visitors to a *marae* pass through the main gateway directly on to the *marae* proper, across which they face the meeting house (*whare* hui or, less commonly today, *whare runanga*). The dining-hall (*whare kai*) is usually placed to one side, with the 'kitchen' (properly a cook house) attached or in an adjoining building, and toilet blocks to the rear. Some *marae* also include a cemetery, church, and sports' grounds. (1967: 175)

As in former times, *marae* proper and meeting-house are extensions one of the other. The *marae* proper is used for welcoming guests and speech-making in the day-time and fine weather, the meeting-house for sleeping in and for speech-making at night and when the weather is bad. Because of their traditional origin and functions, *marae* proper and meeting-house are regarded as *tapu*, and emphasis is placed on the observance of traditional forms. Welcomes and speech-making follow basically traditional patterns. ... Water from the roof of the meeting-house (symbolically the head of the ancestor) is not used but drained back into the ground. Most tribes limit eating andsmoking in the meeting-house: some ban them altogether. In contrast, the dining-hall is an utilitarian building without *tapu*. (1967: 176)

Of all the gatherings held at a *marae*, the *tangihanga* is the most important. Wherever possible, Maoris take their dead to a *marae* (preferably an ancestral one) to 'lie in state' for two or three days, while relatives and friends come 'to pay their respects'. (1967: 177)

The *marae* also serves an important function as a community forum. Whenever Maoris gather at a *marae*, whether for purely local meetings or for *hui*, there is always plenty of speech-making. (1967: 178) residents view Broughton Park as a bounded place, a leafy middle-class 'island' surrounded by a sea of 'no-go' areas. (Valins, 1999: 274)

In a most graphic example, he also documents the politics surrounding proposals for the construction of a communal *eruv* or space where ortho-dox Jews may be exempt from the Jewish law forbidding the carrying of objects on the Sabbath. Some opponents to the proposal included secular Jews who were concerned that such a move would spatially stigmatize the area and create a ghetto that could increase anti-Semitic sentiments. Other orthodox Jews feared 'the social consequences of easing Sabbath restric-tions so that by allowing people the freedom to move with fewer social and spatial constraints, the sanctity of the day [would be] diminished and infractions of Jewish law encouraged' (Valins, 1999: 259). In each case, ethnicity and space were closely entwined, influencing how Jews viewed their neighbourhood and behaved within that space.

At an even smaller scale, Pulvirenti (2000) reminds us that different ethnic groups will establish different meaningful socio-spatial relations associated with housing. Taking the example of postwar Italian immigrants in Melbourne, she shows how the social and political history of these migrants enhanced their value of *sistemazione* or the achievement of 'settlement, a place, a future' (2000: 239). She argues that since migrants had left great poverty in Italy and an extremely low and corrupt quality of life, their material and moral interest in *sistemazione* as an expression of self-sufficiency and survival shaped both their approach to work and their spatial negotiation of the urban environment of Melbourne.

> You leave with the hope of finding a little *sistemazione*. That was what brought us to Australia. Without that little bit of hope you couldn't leave. (Flavia, Transcript 10)
>
> *Sistemazione* means that, for example, when you get your money you make yourself a house with everything inside that is necessary. Clothing is [also] necessary, or you put your money in the bank to think about the future, either an illness, or unemployment or your old age, so that you can say, 'I won't have to go here in search of $1 ... or having to beg the government for something, or to not be able to eat well'. That's a *sistemazione* where you can say 'I have a little money in the bank, I have my house *sistemaziona*. The house is full of everything so that I don't need anything else. It's nice and clean, in order'. That is a *sistemazione* for the future. ... So I am *sistemato* because I don't need anything. (Franco, Transcript 14). (Pulvirenti, 2000: 243, 244)

Noting that Italian migrants enjoyed the Australian period of high postwar employment (1950s and 1960s), she shows how the desire to achieve *sistemazione* was achieved and satisfied through home ownership in key inner-city suburbs – 'Italian-Australians having the highest outright home-ownership rates of any birthplace group' (Pulvirenti, 2000: 237). Migration to Australia and establishing a home were core ways in which Italians could recognize and achieve their own values of security and settlement.

An additional focus in social accounts and critiques of ethnicity include those works that address indigenous populations, who through colonial struggles are decimated or at least marginalized. For instance, Dunn et al. (1995) show how the commercial construction of a new tourism-focused identity for Newcastle (NSW, Australia) omitted many social groups and histories, including the presence and history of an aboriginal population. They document how the Awabakal (Awabagal) indigenous people had developed an effective land and marine economy and society in the coastal area between what colonizers called Lake Macquarie and the Hunter River. Although the Awabakal people were rapidly overrun and dispossessed of their land in the early nineteenth century, the area was shaped by their hunting and gathering lifestyle which included traces such as stone weirs or fish traps in the rivers, notched trees and fire-based vegetation changes. Yet Dunn et al. show that in the *Best City* invention of a post-industrial urban identity for Newcastle, 'an embarrassing Aboriginal presence [and history was] ... ignored and omitted from the public presentation of the new identity' (Dunn et al., 1995: 163). This type of work links directly with the third approach to geographies of race and ethnicity, namely studies of racialization. These works focus on the ways dominant social groups construct ideas of race and ethnicity in order to ignore, silence or control the culture, social relations and/or spaces associated with ethnic minorities.

5.3.3 Processes of racialization and the discursive constructions of race and ethnicity

A growing number of recent geographies have engaged with the discursive and power relations that shape knowledge and experience of race and ethnicity. Drawing on the 'cultural turn' and postmodern and poststructural interests in difference and **discourse**, these geographers have begun to document processes of racialization that have produced racial categories and have racialized whole landscapes.

An important early example of this type of work includes Anderson's (1988) account of how a racial category can be socially constructed. Taking the case of the Chinese in Vancouver (Canada), she analysed how spatial and institutional constructions of 'Chinese' as a racial category of 'outsiders' were established and maintained over a hundred years (1880–1980) – see also Part V. She argued that race is socially constructed in historically-specific conditions and that more attention should be paid to 'the process by which racial categories are themselves constructed and transmitted' (1988: 128). Anderson showed how the Chinese population was constructed as an 'inferior race' and how institutional racism through the Provincial Government of British Columbia ensured Chinese were blocked from voting in elections, entering most professions, working in public or government-assisted projects, and from buying or leasing Crown land. She then identified how racialization was articulated spatially, tracking the changing identity of Vancouver's 'Chinatown' as an emerging evil (1880s), an opium den (1920s), a slum to be rebuilt (1960s), and an ethnic neighbourhood (1970s). She concluded:

> During the evolution of white European domination in Vancouver, British Columbia, conceptions of a 'Chinese race' became inscribed in institutional practice and reconstituted through the locality known, and produced, as 'Chinatown'. ... The space of knowledge called 'Chinatown' grew out of, and came to structure, a politically divisive system of racial discourse that justified domination over people of Chinese origin. (Anderson, 1988: 145, 146)

In a different context, Berg and Kearns (1996) also highlight the significance of spatial and discursive constructions of 'race' in the landscape of New Zealand. Completing a discourse analysis of place names in New Zealand, they show that the colonial and nationalistic projects of naming and mapping a New Zealand territory involved the construction of predominantly English (and masculine) norms and names. The naming of places in this way Anglicized the landscape and constituted part of what Durie (1987: 80–1) calls 'this driving need that permeates our western society, to own, possess and dominate the landscape'. These place names overwrote the names, language and significance of places to the *tangata whenua* ('people of the land').

> [P]lace-names are important signifiers of meaning, providing symbolic identity to people, place, and landscape. Identity, in this sense is not given. Rather, it is 'made' in the contested process of cultural (re)production. ... Place-names, and the maps used to present them, are the outcome of the appropriation of symbolic production by hegemonic groups, who impose their specific identity norms across all social groups. (Berg and Kearns, 1996: 118, 119)

Attention to discursive racism of this kind allows the possibility of resistance and oppositional discourses. In this case, Berg and Kearns document the recent politics of contesting place names through the New Zealand Geographic Board and ways in which these names are pronounced. These processes form an important avenue for Maori to attempt to reconstruct and reclaim identity and landscape through Maori place names (although it is important to note the processes and outcomes have varied – Kearns and Berg, 2002).

Both of these works illustrate the discursive processes of racialization – whether this affects a particular population (as in Anderson's study) or a landscape and nation (as in Berg and Kearn's study). Together with Maddrell's analysis of geography curriculum (discussed in section 5.2), these works indicate how explicit and aggressive racism in the case of colonial and imperial histories may be overlain with discursive and cultural practices that differentiate people by race or ethnicity and then attach powerful assumptions and conclusions to these categories. These studies have been important for two reasons: first, by showing that race is a constructed category, 'whiteness' and the social construction of a hegemonic position for white ethnicities has been exposed for closer scrutiny; second, by highlighting the institutional and discursive processes by which race and ethnicity are constructed, alternative possibilities for identity and political struggle are suggested. Each of these developments constitute part of contemporary geographies of race and ethnicity, and they point to fruitful connections that are being made with and across other types of geography, as will be shown in the final section of this chapter.

5.4 CONCLUSIONS, CHALLENGES AND CONNECTIONS

Geographies of race and ethnicity have developed into a rich field of social and cultural geography. This chapter has introduced only some of the diverse themes and approaches being taken in such work. The three approaches outlined in the previous section show the complementary significance of quantifying, interpreting and discursively critiquing the way race and ethnicity are concentrated, negotiated and constructed through various social systems and spaces. In each case, race and/or ethnicity and space have been shown as mutually constituted. Ethnic diversity has been mapped or read across different spaces. The construction, experience and value of different spaces and places has been interpreted and/or deconstructed as imbued with various cultural and ethnic identities or qualities. While it could be argued that these geographies have become more nuanced and critical over time, a number of challenges remain.

First, geographies of race and ethnicity, irrespective of approach, have increasingly recognized the complexity of racialized space and societies. This chapter has shown that contemporary statistical studies have been important for acknowledging the heterogeneity of ethnic minorities and the socio-economic conditions they face, but these studies have also shown that a huge diversity of locality and ethnicity effects (Byrne, 1998) underpin the heterogeneity quantified. These geographies, together with qualitative studies of different ethnicities, have recognized that racialized or ethnic differences are also transected by other axes of difference, such as class, gender and age (Byrne, 1998; Peake, 1993; Watt, 1998). For instance, Robinson's (1993) study of different internal migration patterns of white, West Indian and Pakistani populations in England and Wales also recorded the influence of gender – showing how experiences of men and women differed greatly between the Pakistani and West Indian groups. Likewise Pulvirenti (2000) has shown that while Italian notions of *sistemazione* affected the work and residential decisions of Australian migrants, men and women differently contributed to the achievement of *sistemazione* through their homes. In many cases, the multiple differences become a cumulative disadvantage as Watt shows for Asian and black young people negotiating the white landscape and society of South East England (see also Gilbert, 1998 for the combined effect of gender, poverty and 'race'). However, these differences cannot be seen in an 'additive sense' of double and triple disadvantage for they intersect and mutually constitute each other (Jackson and Penrose, 1994: 18).

A second challenge that has been approached since the 1990s stems from these geographies of multiple difference. It involves the recognition of ***whiteness*** as an equally constructed ethnicity. This has been an important acknowledgement for 'dominant groups still manage to efface their own actions by implying that they are somehow outside the process of definition' (Jackson and Penrose, 1994: 18). Just as Anderson (1988) argued that Canadian 'Chinese' were a socially constructed racial category, many geographers have begun to analyse critically, and in some cases deconstruct, the often implicit hegemony of whiteness that shapes so many western societies and landscapes (Berg and Kearns, 1996; Jackson, 1998; Maddrell, 1998; Watt, 1998; Zavella, 2000). For instance, works by Gilbert (1998) and Watt (1998) are important for showing how poor African-American women or Asian and black young people negotiate employment, childcare and landscapes that are predominantly white or provide more opportunities for white populations. Similarly, discursive works such as Berg and Kearns (1996) and Maddrell (1998) provide important critiques of white/Anglocentric knowledge. Further deconstructions of whiteness, and the multiple ways it is constituted, should enhance geographies of race and ethnicity, encouraging us to see these as embedded in wider political, economic and philosophical histories and struggles (see the section on Anderson in Part V and Anderson (forthcoming)).

Third, and in a complementary fashion, geographies of race and ethnicity face the challenge of practice as questions are raised about the position of the researchers and of how the geographies are conducted and re-presented. Critiques of geographic practice, as culturally confined to a predominantly white, middle-class, scientific knowledge have provided new points of departure. As in feminist and development geographies, attention is increasingly given to the situated nature of the knowledge geographers create. Driver (1992) and Maddrell (1998) remind us that geographical knowledge has historically been instrumental in colonial and imperial regimes. But contemporary geographies have also become increasingly conscious of the **positionality** of the researchers and the knowledge they are constructing (Jackson and Penrose, 1994; Pratt, 2000b). Sometimes, geographers have even challenged each other over the practice of their geography and the uncritical assumptions and traditions they bring into their work (Jacobs, 1994; Stokes, 1987). An example of this work is provided in Jacobs' (1994) critique of Davis and Prescott's (1992) geography of 'Aboriginal frontier and boundaries' as a problematic mapping and judicial spatial project. In New Zealand, Stokes has been an important figure for challenging Pakeha geography, arguing that:

> Māori geography is not something that is learned only from the written word. ... Māori geography is another way of viewing the world, another dimension, another perspective on New Zealand geography. But – kia tupato, Pakeha. Be careful Pakeha. Tread warily. This is not your history or geography. Do not expect all to be revealed to you. ... You must show respect for the **tapu** of knowledge. Do not expect that because you are an academic or experienced researcher in the pakeha world that all this will come easily to you. Your degrees, your experience, and list of publications, may be more of a hindrance than help. (Stokes, 1987: 121)

With the increasing importance of discourse in poststructural approaches, some contemporary geographies in New Zealand note that the values and knowledge beyond traditional, white, written academic practice will enrich geographies of race and ethnicity (Teariki, 1992; Teariki et al., 1992). McClean et al. (1997) provide an important example, illustrating how knowledge can be co-created with Maori. It is this text that the students cited at the beginning of this chapter drew upon in concluding their seminar. Of all the reading and thinking they did concerning geographies of ethnicity, their most significant learning came in their recognition that geography could be enormously enriched by respectfully considering social and intellectual practices and values beyond the mainstream social science practices they were learning in the majority of their degrees (see Box 5.4).

Reflecting on geography in this way highlights the politics of research and knowledge constructions and complements the final challenge and connection that can be seen in contemporary geographies of race and ethnicity, namely action. Some recent geographies have focused on the actions and goals of groups who are strategically identifying their ethnicity. As shown in section 5.3.2, Valins's (1999) study of Jews in Manchester is one of example, where ethnicity and identity are closely interwoven in active ways that enable people to create personal and collective politics. Other

---

Box 5.4 Maori challenges to the practice of geography

McClean et al. (1997) argue that Spivak's (1996) notion of a 'responsibility structure' should guide the creation of responsible geographies in 'Aotearoa'/New Zealand. Such a structure would:

- Be prepared to co-create knowledge with Māori.
- Ensure appropriate spaces and rituals for negotiation were arranged (e.g. through a *hui*).
- Be prepared to negotiate 'border-crossings' between cultures and discourses and the positions that are staked within them.
- Negotiate different interests and expectations regarding output/written reports.
- Hold a commitment to *aroha* – an untranslatable notion that entwines ideas of love, care, responsibility and preparedness to be accountable in a form of unity that does not force sameness.

L.T. Smith (1999) identifies culturally specific ideas associated with Kaupapa Māori practices. The codes of conduct she has published can guide geographies in New Zealand, and may stimulate other appropriate codes for different cultural and ethnicity-focused research. The code involves:

1. *Aroha ki te tangata* (a respect for people).
2. *Kanohi kitea* (the seen face, i.e. present yourself to people face to face).
3. *Titiro, whakarongo … korero* (look, listen … speak).
4. *Manaaki ki te tangata* (share and host people, be generous).
5. *Kia tupato* (be cautious).
6. *Kaua e takahia te mana o te tangata* (do not trample over the *mana* of people).
7. *Kaua e mahaki* (don't flaunt your knowledge). (1999: 119–20)

---

examples include the work of Dwyer (1998), documenting Muslin young women's personal choices in identity and dress, of Pratt (1999a, 1999b) interpreting Filipina domestic workers negotiate different work conditions, institutions and spaces, and C. Gibson (1998) analysing Australian aboriginal music as a medium for indigenous self-determination. These studies are discussed more fully in Chapters 8 and 10, but they signal exciting new directions where identity and politics are negotiated so that racialized ethnic groups can mobilize personal or collective actions, actions which can shape lives and spaces that enable their cultures and preferences for self-determination, or at least make them visible.

SUMMARY

- Geographies of race and ethnicity focus on groups and individuals who can be identified and differentiated by their racialized or ethnic affiliation.
- Racial identification has concentrated on physiological and phenotype differences while ethnic identification has concentrated on a sense of shared ancestry, values and cultural practices.
- Some geographers have statistically analysed and mapped spatial concentrations and socio-economic patterns of racialized or ethnic groups in terms of dimensions such as migration, housing and employment.
- Other geographers have gathered and interpreted accounts of different groups' experiences of space and place – as they are influenced or mutually constituted by ethnicity.
- Still other geographers have critiqued the way processes of racialization and racism have been constructed discursively and spatially, through such examples as the design of educational curriculum and the naming of colonial landscapes.
- Today, geographies of race and ethnicity intersect with, and are challenged (or enhanced) by, other geographies addressing multiple differences (e.g. intersections of race, gender and class) as well as those focusing more particularly on identity formation, activism and politics.

Suggested reading

A good introduction to questions of race and ethnicity is provided in Jackson's (1992) book *Maps of Meaning*. While many works have been published since then, his sixth chapter on 'Languages of racism' explains and illustrates the way race has been constructed over time in a variety of settings. More detailed research-based examples of contrasting geographies of race and ethnicity include:

- Rees et al.'s (1995) quantitative analysis of the heterogeneity of socio-economic circumstances of different ethnic groups in two British cities;
- Pulvirenti's (2000) qualitative narrative of how Italian ethnicity, including gender relations and notions of *sistemazione*, affected post-Second World War Australian migrants housing ambitions and home-ownership patterns;
- Berg and Kearns's (1996) critical reading of how the landscape of New Zealand was colonized through the use of (and struggle over) place names;
- Jacobs' (1994) informative interpretation of how a key site of conflict between an Australian aboriginal group and a mining company involved economic,

geographic and sacred discourses that were entwined in the struggle over this land; and

- Anderson's (2000b) wider reflections on how ideas of race have been and will continue to be influential and how people and institutions construct national (or post-national) identities.

6 Sexuality

6.1 INTRODUCTION

We negotiate *sexuality* every day. Casual conversations in a workplace, a walk on the beach with your partner, or time out with friends in town – each of these activities include negotiating the social meanings and values about sexuality that are associated with our bodies and actions and the spaces we are using or passing through. One student recently summed this up by reflecting on geographies of sexuality and their own experiences:

> I've never thought about this before. Reading these studies changes everything. What I saw as just a place, like a pub or The Octagon [an open-air plaza in the centre of Dunedin], are no longer neutral spaces. They are *straight*. If you are gay it would be so hard to be 'out' with a partner there. I never thought what it must be like to walk through spaces like this. (Student Log, 1998)

Another student wrote about the more indirect interactions and negotiations that might follow from their academic transcript:

> In the transcript of my academic studies the first thing in the list before all my geography courses is a paper under Women's Studies, named 'Lesbian and queer studies'. That is the only Women's Studies paper I have taken, and for a non-geographer it apparently has no connection to my other papers. In that case, is it clear evidence of my sexual identity? And further, if I show my academic record in a job interview to a middle-aged heterosexual (homophobic?) male, what is likely to happen? (Student Log, 2002)

Each of these examples illustrates that sexuality is central to our daily lives and often our future choices or opportunities. It affects the spaces in which we live and the social relations and judgements we may face.

Geographies of sexuality have been the most recent core category of social difference to be widely acknowledged in social geography. Since the 1970s, but particularly the latter part of the 1980s and 1990s, geographers have shown how sexuality and space mutually affect each other. Gradually geographers documented how space, and specific places, are sexualized through different processes that convey dominant and alternative meanings and norms about sexuality. And they also showed how spaces (and the socio-cultural relations that shape them) also affect our sexual identities and the ways we experience and negotiate these spaces – whether it be our own bodies, or the spaces of home, work, leisure sites and so-called 'public spaces'. But geographies of sexuality have not been homogeneous. Instead, this chapter shows that sexuality has been understood in a variety of ways (section 6.2) and that early geographies sought to map or explain sexuality in relatively definitive ways (section 6.3) while more recent works have challenged many early assumptions and have shown how sexuality is constructed through both material and cultural processes as a fluid rather than a fixed social category (section 6.4). The biography in Part V describes the work of Larry Knopp to complement this chapter, illustrating many of these same themes through his work.

6.2 DEFINING AND THEORIZING SEXUALITY

Geographies focusing on sexuality have a relatively short history compared to those of other differences such as race and ethnicity. While many traditional geographies are now being acknowledged as **heteronormative**, and heterosexual assumptions implicitly incorporated in much research are being critiqued, sexuality as an explicit subject of social geography did not emerge until the 1970s. Wider societal changes, including the impact of sexual and gay liberation movements, increased the acceptance of sexuality as a legitimate subject of social inquiry. However, in geography, studies of sexuality did not become well recognized until the 1990s.

The key ways in which sexuality has been approached in social geography are summarized in Table 6.1. This table illustrates how sexuality has been conceptualized using the different approaches already identified in previous chapters. Early studies focused on sexuality as a behaviour (or set of sexual acts) that generated map-able activities. Results showed the location and organization of activities such as prostitution. In this kind of work, sexuality was predominantly defined through specific sexual behaviour and the economic, legal and spatial relations that are directly related to these acts. Geographies of this kind thus parallel other more quantitative social geographies of the time. They identified and classified the spatial patterns involved. Sex was one more social act that generated different locational features: activity spaces, nodes and spatial hierarchies (details are provided in section 6.3).

TABLE 6.1   CONTRASTING GEOGRAPHIES OF SEXUALITY: APPROACH AND FOCI

Type of geography: theoretical approach	Foci
Quantitative and humanist mapping and accounts	Sexuality as *sexual behaviour*. Studies focus on 'problems': prostitutions and deviant areas.
Marxist and feminist critiques	Sexuality is understood in terms of unequal *power relations* (through the dominance of capitalism, and/or, patriarchy, and/or heterosexuality).
	Sex can be understood as a commodity in capitalism. Studies focus on the production and consumption of different sexual relations in different parts of the city.
	Sexuality and gender are closely connected in *heteropatriarchal* relations. Studies analyse how women are consequently disadvantaged in society and space.
Poststructural readings and queer alternatives	Sexuality is read as a *dynamic social and cultural construct* rather than a stable or fixed category.
	Studies focus on how meanings and choices surrounding sexuality are constructed socially and spatially and are predominantly *heteronormative*.
	Sexuality as a broad dimension of life is performed and can be mobilized through strategic identities for political purposes.
	Geographies can be written that challenge the heteronormative assumptions of previous theory.

In contrast to this relatively descriptive way of locating sexuality in space, feminist and Marxist accounts have taken more critical and explanatory approaches to the definition and study of sexuality. Rather than treating sexuality as a set of sexual acts that can be plotted and mapped, radical approaches sought to define sexuality in terms of unequal power relations. Radical approaches built critical explanations of these relations and how they differentially affect various groups of people. Men were identified to dominate and be advantaged sexually compared to women, and heterosexuality was acknowledged as a dominating and powerful sexual culture, compared to homosexuality, bisexuality and transsexuality.

Feminist approaches through the 1980s understood sexuality as part of gender and sexual relations that privileged most men and oppressed, or at least disadvantaged, women. Theoretical distinctions between radical and socialist feminists produced contrasting emphases on sexual and economic causes of inequality and oppression of women. Radical feminism supported geographies that tended to focus on intersections between sexuality, patriarchy and experiences of being a woman. This included the uneven experiences of women with dependent children, the alternative spaces lesbian women might create in strategic resistance to patriarchal relations, or the everyday experiences of fear of violence in public space that women might negotiate. Socialist feminism supported geographies focusing on women's experience of gender and class within capitalist societies and economies (compare, for instance, Johnson, 1987 and McDowell, 1986 in response to Foord and Gregson, 1986). Beyond these differences feminist geographies began to critique the previously naturalized assumptions about heterosexuality underlying notions of 'family' and 'work'. Not only were sexual reproduction and sexual relations gradually seen to affect women's experience of family and access to paid employment, but also they were understood to underpin explanations of **patriarchy** and the generic inequalities between men and women. Moreover, feminist geographers understood sexuality and gender as a series of unequal relations dominated by **heteropatriarchal** assumptions that would have spatial dimensions, including access to housing, negotiation of public space and experiences of safety and violence.

Marxist geographies of sexuality also focused on unequal power relations, but concentrated on the way sex could be understood as a commodity in capitalism that might be 'produced' and 'consumed'. The commodification of sex and the understanding of sexual relations operating within capitalism also enabled geographers to concentrate on the different places and spaces in which sexuality was differentially produced and consumed. Conceptualizing sexuality in this way aided theories of how sexuality is complexly involved in the reproduction of the economic and social character of capitalism in different (especially urban) spaces and contexts (see section 6.3.2).

While Marxist and feminist approaches to sexuality continue to define aspects of the concept and the way it is studied in contemporary social geographies, newer postmodern, poststructural and **queer** conceptualizations have extended notions of sexuality. Such conceptualizations highlight its fluidity or instability and its potency as a political dimension of personal and collective social life. Postmodern and poststructural attention to difference, diversity and the cultural construction of meanings and identities has provided new avenues for understanding. Sexuality is read as a diverse set of sexual preferences and identities which are culturally and socially assembled or constructed through discourses, spaces and the relations between people. For instance, discourses of family or desire will construct subject positions and behaviours that convey various sexual identities and values attached to individuals or groups. Some meanings and interests will be legitimized while others are marginalized (see Box 6.1). Bodies and other spaces (such as streets, bars or workplaces) may be read as sexualized

---

Box 6.1 Public toilets in a botanical garden

Many places within our everyday lives are valued and carry meanings in a diversity of ways. A set of toilets in a botanical garden is one such place. In 2000, the toilets in the Dunedin gardens (above) became a contested space as different groups were identified as having contrasting and conflicting views about the function of this space. Was this a local government amenity for the public? Was it a 'family space'? Was it a location for informal and intermittent gay male sex? Was it a crime site? In 2000, the daily newspaper ran a front-page news story. In this case, one set of meanings and uses of the space are conveyed as sacrosanct and another as inappropriate and criminal:

> The Dunedin Botanic Garden is becoming a magnet for men looking for casual homosexual activity but the police warn the behaviour is not appropriate for a public area used by families. ... Sen Const Steel said the behaviour was not considered appropriate for an area used by families and young children, and police intended taking a hard line with offenders. (Page, 2000: 1)

Source of photograph: Panelli, private collection.

(usually with a dominant heterosexual theme). Thus geographies of sexuality drawing on poststructural theory focus on the socio-cultural construction of sexual interests, activities and identities. Attention to difference ensures that the diversity of sexualities is studied – especially drawing on queer theory.

Just as poststructural geographies have challenged many of the meta-narratives and scientific assumptions of past work, queer geographies have drawn on queer theory to provide alternative notions of sexuality and establish critiques of mainstream geography as *heteronormative*. Queer approaches to sexuality destabilize sexual categories and theories, suggesting that our sexuality is neither necessarily fixed nor stable through time and space. Queer work also highlights the *performativity* and politics of sexual identities. Geographers drawing on this perspective show how spaces and places are mobilized through politically-infused actions such as creating lesbian archives and gay pride marches (see section 6.4 for details).

In sum, while sexuality might be approached individually and socially as both physical and socio-cultural dimensions of sexual desire, and the wider expression of different sexual identities, social geographers have drawn on the heritage of contrasting approaches to study this form of social difference. Sexuality has been variously understood and studied as:

- sexual behaviour that produces specific spatial patterns;
- sexual relations that are uneven and frequently advantage heterosexual males;
- a commodity (as sex) that can be economically and culturally produced and even incorporated into other spatial processes in capitalist societies; and
- a set of discursive identities that can be performed, shifted, mixed and sometimes strategically assembled for political purposes and that challenge the dominance of heteronormative ideas and sexual identities.

These different conceptualizations of sexuality are apparent in the following two sections that present contrasting geographies of sexuality.

6.3 EARLY GEOGRAPHIES: MAPPING AND EXPLAINING PATTERNS OF SEXUALITY

6.3.1 Locating sexuality in space

Early geographies of sexuality should be considered in the context of the modernist impulses and practices that shaped much of postwar quantitative geography (see Chapter 2, section 2.3). The quest to map, classify and consider the management of social problems was no less relevant in early geographies of sexuality. Underpinning this work was an assumption that 'private' heterosexual relations were acceptable and normal, while prostitution or homosexuality were identified as problematic. Richard Symanski's (1974, 1981) research is a key example of this type of work. His studies of prostitution included detailed mapping and descriptions of the location, patterns and processes associated with prostitution in both the United Kingdom and USA.

Symanski used different scales of analysis as he documented activity spaces and hierarchies of settlements and brothels in different settings.

Interest in classifying the spatial character of social differences enabled Symanski to draw on the then current notions of social class and settlement hierarchies. He noted small-scale, low-class brothels to 'inaccessible rural areas' and larger, 'better' or 'plusher' brothels with 'high class girls' located in larger settlements within a state-wide settlement hierarchy (see Figure 6.1). Symanski successfully demonstrated a geography of prostitution at numerous scales. For instance, at the state level in the USA, Symanski classified counties and centres by the legal status of prostitution and their brothel numbers, as well as by numbers of prostitutes (measured by arrests) and types of clientele. At finer scales, he mapped street and bar-scale patterns of prostitution, and analysed the city-wide spatial occurrence and changes in prostitution following amendments to police policies and tactics.

Other early spatial interests, influenced by the Chicago School of social ecology, have resulted in maps and/or descriptions of the location and concentration of gay and lesbian populations. Loyd and Rowntree (1978) and Weightman (1981) used social and spatial notions of 'community' to document the existence of gay communities, while sociologists developed some spatial accounts in their studies of gay neighbourhoods and lesbian ghettos (Castells, 1983; Ettorre, 1978, 1980). These studies provided important early documentation of sexuality as a set of social and spatial relations. Space as differentiated social territory was a vital part of these geographies and sociologies. In the case of housing and lesbian activities, Ettorre described a London 'lesbian ghetto':

> The ghetto lasted approximately six years (1971–77). ... [T]he ghetto involved approximately 50 lesbian feminists ... as residents of the area and approximately 150 feminists as non-residents. Within the radius of half a mile, the lesbian feminist ghetto emerged, grew and developed as the centre for lesbian feminist activity not only within the area of Lambeth but also within the urban context of London. Particular spatial arrangements and consumption practices evolved as did localized, yet developing feminist practice. Within this agglomeration (consumptive unit), most of the lesbian feminists occupied housing facilities in four blocks of flats situated on two city streets. The streets were parallel to each other and one city block at a distance from each other. (Ettorre, 1980: 514–15)

Major differences in interpretation did occur in these early works. While Ettorre documented lesbians' ability to concentrate their interests spatially, Castells argued that gay males were more likely than lesbians to form gay territories because of their more advantageous socio-economic positions and their greater inclination to hold spatial or territorial aspirations compared to lesbian women. Over time, however, lesbian studies have affirmed Ettorre's findings and have continued to modify and/or challenge Castells' contention, showing the variety of ways lesbians create spatial networks and spaces (Adler and Brenner, 1992; Peake, 1993). Despite

(a)

(b)

FIGURE 6.1   BROTHEL DIFFERENCES BY CLASS AND LOCATION

Source: Symanski (1974: Figures 6 and 7).

these debates, across the different accounts, space has been presented as a crucial medium and resource by which gay and lesbian populations could concentrate housing and lifestyle interests through their social networks and spatial choices. Moreover, all of these early works were important for making sexuality and sexual differences visible in the physical land-scapes of the states, cities and streets of western societies. They also laid a foundation of academic literature and dialogue for later geographers who sought to provide broader explanations beyond the maps and case descriptions initially published.

6.3.2 Feminist and Marxist explanations

Feminist and Marxist geographies of sexuality each used critical approaches to develop theories of why and how cities included sexual landscapes where uneven gender and sexual relations resulted from capitalist and patriarchal social systems. However, this literature blurs and defeats labels since, to a large extent, Marxist analyses drew on a great deal of feminist literature and much feminist geography reflected the influence of Marxist theory through the expression of socialist feminist works.

Feminist geographers, working in interdisciplinary ways with other feminist scholars, came to include sexuality in accounts of urban life. Drawing on other (especially economic) writings, feminist researchers began to incorporate acknowledgements of the heterosexual and patriarchal assumptions underlying the organization of urban space and people's access to housing, finance and work. Western cities were noted as social and economic spaces where heterosexual and gender-biased meanings and processes affected where men and women worked, their positions in relation to the social reproduction of capitalism, their roles as consumers, and their differing access to spaces and services such as housing options (Bowlby et al., 1982, 1986; England, 1991; McDowell, 1983; Tivers, 1978; Watson, 1988; Wekerle, 1984). In the case of housing in the USA, England showed the post-Second World War government policies were directly geared to nuclear, heterosexual families:

> The Federal Housing Administration used zoning to protect entire sub-divisions of middle-class, single-family, owner-occupied houses. Clearly, the suburbs were oriented toward 'traditional' nuclear family life and were seen as an ideal environment in which to raise children. ... By defining 'family' in its limited sense, moreover, sharing a house with unrelated individuals ... was prohibited. Clearly zoning, then as now, limits the opportunities of both non-traditional and non-heterosexual families. (England, 1991: 138)

Similar arrangements privileging nuclear heterosexual family units were also common in Australian state housing. Watson (1988) showed the housing industry as a set of institutions and practices that maintained a dominant order of gender and sexuality. These hetero-sexist practices not only affected who could gain housing, but also who could gain housing finance – with various finance companies requiring men to stand as guarantors for women, or providing differing financial arrangements to women who were prepared to be sterilized.

Other feminist geographies have also been written specifically to increase the visibility and voice of lesbian women in geographic research and literature. As a feminist practice of writing women *into* geography, these early lesbian geographies were important for enabling one group of particularly marginalized and invisible women to be 'seen' and read in geography literature. Gill Valentine's (1993a, 1993b) accounts of lesbian experiences are the most well known of this type. She argues that heterosexuality and patriarchy intertwine so that gay and lesbian identities involve negotiating and challenging the dominant norms of family structure, sexual behaviour and gender identities regarding masculinity and femininity (1993a: 298). The 'everyday' lives and spatial experiences of these negotiations are clearly documented and show the range of spaces and social relations lesbians have to negotiate (see Table 6.2).

Parallel to feminist explanations of sexuality and space, Marxist geographers have sought to critique and explain sexual inequalities, including different women's and men's experiences of space, work, resources and social life. Drawing on traditional Marxist frames of analysis, these works have set sexuality within a wider critique of capitalism and class processes. Knopp's research into sexuality and the dynamics of urban social processes such as gentrification illustrates this approach (see Part V and Knopp, 1987, 1990b, 1992). Note that Knopp is another geographer who has worked through a range of analytic resources provided by different theoretical approaches (as exemplified by Cloke and Pratt in Chapter 2). A biography of his work (Part V) shows that while his early works, discussed here, are greatly influenced by critiques of capitalism, his later writings (1995, 1998; Knopp and Brown, 2002) draw more from contemporary feminist, postmodernist and poststructuralist thought. In the early 1990s, Knopp employed a Marxist approach to argue that:

> [S]exuality is implicated in the spatial constitution of [capitalist] society and simultaneously, ... space and place are implicated in the constitution of sexual practices and sexual identity. ... [T]he spatial dynamics of production, consumption and exchange (including particularly class struggle) under various phases of capitalist development [are linked] to the range of possibilities for sexual practice and sexual identity formation that confront people in their daily lives, as well as to people's struggles, individually and collectively, to negotiate and shape these possibilities. (1992: 652)

Illustrating his argument with different time periods, Knopp argues that the rise of industrial capitalism spatially and economically separated production into private/domestic and public/exchange forms that resulted in gender divisions and contrasting sexual opportunities or expectations. Knopp notes that early industrial capitalism and women's economic dependence on men meant that independent lesbian lives and homes were beyond the reach of all but the most affluent classes while, for men, 'sex (like nearly everything else) [became] something very close to a commodity. It was something to be objectified, traded for, and consumed' (1992: 659).

TABLE 6.2   LESBIAN EXPERIENCES AND NEGOTIATIONS OF 'EVERYDAY SPACES'

Everyday space	Lesbian interviewees' personal accounts, examples and descriptions	Sources from Valentine's work
Home	My first house in Kerrison Avenue – we came out one morning and 'lesbians live here' was written on the footpath of the house.	(1993a: 398)
	I've got several books with the word lesbian in the title. When Karen [colleague] came round, Monica and Anna were in the lounge, so I took her upstairs. I went to make a cup of tea and came back and she was looking at my books and I thought oh God, I've forgotten to hide them.	(1993b: 242)
Work	In the job I'm in they all talk about their men and their husbands and they've had a nice weekend and done this and done that. And I basically can't say anything. And I find it quite difficult because I'm quite an outgoing person. … I was sort of going to a party on Saturday and they were saying 'I bet you meet your Mr Wonderful this weekend'. And I'm going 'I bet I won't'.	(1993a: 403)
	[When I had a job] I didn't like it because I had to dress in a particular way and I just didn't feel at ease in a skirt and make-up, but I used to make the effort.	(1993b: 242)
'Public' space: streets and leisure spaces	I do feel left out. I hate going to straight places. I really hate it. I just feel I don't belong so I prefer not to be there where everyone else is straight, where I'm the odd one out. Like if you go to a restaurant as two women you get a bad table. Yet a bloke and a woman get put in the front of the house.	(1993a: 405)
	We tend not to hold hands on a Saturday in the High Street, as I did with my husband.	(1993b: 242)
	I'm very conscious of being a couple in public in shops and restaurants and things. If I was in a strange town I'd be quite happy about it because I wouldn't bump into anyone, it wouldn't get back to my parents.	(1993b: 243)

With greater access to separate economic independence and inner-city spaces and activities, gay-male opportunities were more numerous. Knopp is also clear, however, about the differences in opportunities and cultures of masculinity that shaped homosexual possibilities within different classes, such as the contrasting lifestyles and consumption options between professional and working-class men.

Knopp's Marxist-informed critique of postwar American urban sexuality cumulatively shows that homes, suburbs, workspaces and public places have been crucially structured around heterosexual and patriarchal capitalism. However, he also shows how later forms of 'postmodern' or 'post-industrial' capitalism, with its attendant de-industrialisation of inner-city neighbourhoods, has provided some opportunities for alternative lifestyle and gentrification processes, especially for certain classes of gay males:

> [T]he presence of an inexpensive, renovatable housing stuck in depressed neighborhoods, which held no particular appeal to any other monied potential market, gave gay people (and again, particularly gay men) the opportunity to develop a territorial *and* economic base for achieving political power and for developing community resources. (Knopp, 1992: 665)

In sum, this section has showed that the geographies of sexuality were initially devoted to mapping and explaining sexual behaviour and relations. The earliest works were most influenced by broader trends in quantitative methods and the practice of geography as a spatial science. Thus the importance of analysing the spatial distribution and hierarchies of sexual behaviour featured in some of the earliest research. While these approaches could be criticized for 'freezing' sexuality in terms of specific acts and moments in space and time, they were nevertheless significant for introducing sexuality as a legitimate social subject within geography, and they were important for introducing examples of how sexuality took on a spatial form. Later critical geographies have sought to provide broad explanations of sexuality based on feminist and Marxist theory. Feminist and Marxist explanations situated sexuality within the spaces and processes of wider social and economic systems. With respective emphases on patriarchal and capitalist relations, sexual meanings and relations were seen to be unevenly differentiated. They argued that heterosexual norms dominate spaces and economic systems of production, reproduction, exchange and consumption, except where specific economic and political opportunities or actions have enabled the acknowledgement or support of alternative sexualities.

6.4 CONTEMPORARY GEOGRAPHIES: MAPPING SEXUAL IDENTITIES, POWER RELATIONS AND QUEER ALTERNATIVES

It is somewhat artificial to draw a boundary between critical Marxist and feminist explanations of sexual geographies and the more contemporary works of the last decade since the critical tone and interests continue in much of the recent work. In fact, geographies of sexuality are a clear example of the palimpsestual character of the discipline discussed in Chapter 2, for early critical geographies have provided a heritage and foundation for recent works by both established 'names' in the field (such as Knopp and Valentine) and newer writers (e.g. Hubbard, 2001; Podmore, 2001).

Contemporary geographies of sexuality extended early approaches but they have also gained further impetus and critical perspectives from the wider developments in postmodern and poststructural thought. These new impulses encouraged attention to difference, the socio-cultural construction of meanings and the politics of uneven social relations and identity formations throughout all facets of human geography. Geographers focusing on sexuality have taken these priorities and developed substantial fields of work on:

- sexual differences (and the fluid, complex and multiple nature of sexuality);
- the socio-cultural processes and effects surrounding what it means to be homo/hetero/inter/bi/transsexual; and
- the contested politics individuals and groups engage with when negotiating predominantly heterosexualized space or the creation of alternative gay, lesbian or other 'queer' identities or spaces.

Many of these geographies have also been influenced by **queer** politics and theory as outlined in section 6.2. The (queer) critical possibilities of broadening and challenging theories of sexuality have been particularly important. Queer theory has supported the analysis and unravelling of sexual certainties and has sustained interest in alternative sexual politics. And, in their own writing of queer geographies (e.g. Costello and Hodge, 1999; Cream, 1995; Johnston, 2002), geographers have created alternatives in the academic arena just as queer activism has mobilized community and street politics to counter the hegemony of heterosexuality.

The resultant geographies can be distinguished in numerous ways. One overview of this recent material is given here using three foci. First, geographers are increasingly reading and critiquing notions of sexual identity, highlighting the diversity, unstable nature and unequal attention and legitimacy given to various sexual categories (section 6.4.1). Second, geographers have also studied the power relations and politics of sexuality (including identity politics), noting how heterosexuality is maintained as a dominant discourse while other political possibilities and resistance can produce different meanings and experiences (section 6.4.2). Third, across and within these two foci, the importance of space is investigated. Geographers have concentrated efforts on the importance of different socio-spatial settings (material, social and symbolic) through which sexuality is established or contested. These range from the scripting and performance of bodies in sexual identity formation, through urban and rural environments that influence sexual meanings and relations, to the spaces that can be claimed or re-constructed in political activities, such as gay-pride marches. Each of these themes – of sexual identity and power relations – and the simultaneous spatial dimensions involved are outlined in the following discussion.

6.4.1 Sexual identities

Contemporary theories of identity and identity politics (the focus of Chapter 7) have been a key feature of recent geographies of sexuality. This contrasts with early geographies of sexuality that had begun to document different sexual preferences and activities by using rather coarse, un-nuanced understandings of prostitutes, lesbians and gay men. Likewise, the importance of constructing gay geographies led at first to an homogenizing focus on the shared or common experiences of lesbians or gay males *vis-à-vis* heterosexuality. Nevertheless, even the early studies by authors such as Knopp (1992) and Valentine (1993a) included reference to the heterogeneity of sexualities (especially differentiation by class). These pieces laid the basis for more detailed work.

Sexuality, as an individual experience and form of identity, has continued to be reported in many studies. Most of these authors have drawn on poststructural (and in some cases feminist) theories to document the way sexuality is lived and **performed** in a variety of ways. Valentine's (1993b) early work had alerted geographers to the multiple identities lesbians had to project through the different aspects and spaces of their lives. As lesbians creating their own homes, as neighbours, shoppers, and pedestrians in predominantly straight streets, as workers in overwhelmingly hetero-sexual workplaces, or as daughters revisiting family homes, lesbian women recounted the dress, behaviour, language and interactions that constituted the various identities they took on in different settings. For instance, Johnston and Valentine have illustrated how lesbians have to perform a variety of identities, choosing clothes and monitoring their bodies in a variety of ways, depending on where they are:

Jackie, a New Zealand lesbian: [I] cover my tattoos up when I go home, especially if mum and dad have company coming over. I do that, it doesn't worry me, that's it.

Hayley, a New Zealand lesbian: [I dress more] conservatively ... kind of straight and less scruffy. (Johnston and Valentine, 1995: 103)

This type of approach has continued (Elwood, 2000; Johnston and Valentine, 1995) and become more complex as it has been joined by other perspectives that counter the earlier 'uncritical all-embracing concept of lesbian and gay identity' (Binnie and Valentine, 1999: 181). Recent studies have also shown the complex use of space in sexual identity formation. For instance, Elwood (2000) and Johnston and Valentine (1995) also have shown how lesbians use home spaces creatively to support their lives. Elwood documented different lesbian strategies in this way:

[Example 1] A bunch of us formed a cooperative household. We'd have gatherings for women looking for lesbian community. We'd invite women who lived alone to be sort of satellites to our family, so they could still live alone, but share a sense of community. ...

[Example 2] The woman upstairs was a lesbian and we formed like an extended family. We did things like take care of each other's kids. The doors to both houses were open and the kids went back and forth. (Elwood, 2000: 19–20)

While home spaces might initially be seen as creative and positive settings for lesbian living (once a person has created their individual living arrangements), they are nevertheless dynamic. Just as individuals monitor themselves and their bodies, they also reflect on and manage their home spaces, depending on who is there and for what purpose:

Having a home of one's own may allow a woman enough control over the space to express her sexuality in the physical environment but it doesn't necessarily guarantee freedom from the prying eyes of parents, relatives and neighbours. Discouraging people from popping in and trying to arrange planned rather than spontaneous visits can buy enough time for the home to be 'prepared' for visitors. Alternatively, visitors can be limited to one or two rooms that are 'produced' for public scrutiny. ... One New Zealander explained that she restricted her parents' movements within her home in order to stop them entering rooms covered with lesbian posters. ... One way to take the tension out of these fraught occasions is to change the performance of the home according to the identity of the visitor. ... some women 'de-dyke' the house completely. (Johnston and Valentine, 1995: 106)

Using queer theory as a means to challenge and subvert fixed notions of identity, contemporary researchers have also highlighted the slippage that can occur between a number of sexual identities – where individuals move between sexualities or gather them into varying combinations (Cream, 1995; Murray, 1995; Wincapaw, 2000). These works have been particularly important for unsettling any spatially stable or socially homogenizing senses of gayness or lesbianism. They have also encouraged the unravelling of the dominant binary of homo/heterosexuality (e.g. Hemmings, 1995). For instance, Wincapaw's (2000) study of sexuality in new hyperspaces of the World Wide Web provides an account of bisexual women participating in a lesbian email-list where women juggle multiple sexualities – expressing lesbian interests and suppressing bisexual ones. Also in this genre of multiple identities, Murray (1995) and Costello and Hodge (1999) show the instability of homo/heterosexual binaries. They document the different ways that 'straight', 'gay' and 'queer' categories can exist in various forms and contexts across (hetero, homo, and other) sexualities – and within single places or sites. Costello and Hodge explain:

> We cannot assume that one's sex, gender and sexuality are fixed. Nor can we ever assume that desire is nomolithic and stable over time. The fact that we are now unsure of what constitutes sex, gender and sexuality produces a fracturing of what can be called heteronormativity. ... Heteronormativity also needs the internal categories of man/woman, heterosexuality/homosexuality, and male/female to remain stable (Jagose, 1996). When the validity of these binaries is challenged, the entire sexual order is unsettled. (Costello and Hodge, 1999: 141)

Despite the way queer geographies can challenge the assumed coherence and stability of sexual categories, there are uneven results and implications from theorizing dynamic and unfixed sexualities. Authors like Murray (1995: 74), Bell (1994) and Munt (1995) acknowledge the politics and unequal attention being given to the interests and accounts concerning different sexualities. Certain sexual lifestyles and experiences are written about more frequently, while others are less common (Murray, 1995). For instance, Bell (1995) and Hemmings (1995) respectively show the more difficult transgressive and perverse positions that unfold in physical sites, legal (public/private) arenas and theoretical (academic) spaces when considering how sadomasochism and bisexuality occur.

A further feature of studies exploring sexual identities involves the greater attention to bodies as sites and spaces where sex is identified. Cream's (1995) analysis of transsexual and intersexed bodies has been an important piece for highlighting the ways gender and sexuality are embodied through medical, biological and social discourses that identify who people are and what their bodies mean. She concludes: 'We need new ways of thinking about nature and biology. ... Corporeal truths are deeply embedded within ideological discourses, and are used to legitimate what people can and cannot do, as well as their place in society' (Cream, 1995: 40).

Beyond the more challenging cases found in Cream's work, other geographies have studied the common ways bodies are identified and sexed. This is negotiated and constructed both through personal decisions and the expectations and surveillance of wider institutions and norms. McDowell's (1995) study of the finance industry and Johnston's (1995, 1996) analysis of body building, show two examples of how discourses of masculinity, femininity and sexuality are circulated throughout the language, spaces, 'rules' and culture of merchant banks and gymnasiums.

Finally, a queer and/or poststructural interest in the connections between bodies, sexual identities, discourses and spaces has been further developed as geographers have begun to consider the broader physical and cultural settings in which people live, such as urban and rural spaces. Urban geographies of sexuality have dominated much of the literature, including the study of how urban societies and spaces provide more diversity and anonymity for the expression of different sexualities. However, recent studies have also documented physical and cultural dimensions of rural environments that provide either opportunities or difficulties for different sexualities. Bell and Valentine's (1995b) review of 'queer country' was a ground-breaking commentary on existing work and geographic themes that could be addressed in establishing rural geographies of sexuality. They recorded the symbolic significance of rurality in gay imagination and cultural materials (novels, poetry, film, etc.), where arcadian ideas of a safe and beautiful countryside and mythical or fantastic/wild landscapes provide a setting for same-sex desire and creative lifestyle. The idealization of rural spaces in gay/lesbian lives and actions has been further recognized by Bell (2000: 549), who notes that 'the rural [is] a symbolic resource in homosexual cultures ... the "constructed countryside" is sexualized or eroticized in the homosexual imaginary'. However, even in these cultural resources, geography is important since the repertoire of identities and landscapes available to people varies in different environments and cultures. For example, the countryside of the United Kingdom provides quite different imaginations from the cowboy and frontier landscapes of the USA or the bush or desert settings of Australia.

Alongside cultural readings of idyllic or utopian rural sexualities, other studies have documented how 'rural gay life' can be extremely difficult and isolating in the face of widespread homophobic and heterosexual attitudes and social relations. Sexual identities and lifestyles that counter the norms of heterosexuality involve different uses of space as well as the threat of abuse or attack. Work in Australia, the USA and the United Kingdom has shown how people differing from the heterosexual expectations of rural life may:

- have to negotiate ostracism or abuse (Roberts, 1995);
- face sexual isolation and the need to use transitory spaces for 'episodic encounters' (Bell and Valentine, 1995b; Kramer, 1995); or
- consider creating alternative, separatist rural spaces for events or communal living beyond heterosexual expectations (Valentine, 1997).

While gay and lesbian geographies of rural spaces have formed important initial developments, it can be seen (as in the wider sexuality literature) that silences and gaps in the literature continue to exist both with more challenging bi/trans/intersex identities and with heterosexuality. Only recently, have geographers begun to interrogate rural environments and societies in terms of the heteronormative cultures that dominate rural communities and discourses (Little, 2002a).

In sum, contemporary geographies focusing on sexual identity have featured the range of multiple and dynamic sexualities that can be understood.

They have also documented the importance of spatial arenas and processes through which bodies, homes, worksites and other environments are sexed. The poststructural and queer theory that has best supported these works also includes/addresses how power intersects with sexuality and sexualized spatial relations. This then forms the focus of the following section.

6.4.2 Power and geographies of sexual politics

Issues of power and politics have been a second common theme in contemporary studies of sexuality. This section sketches out new developments in reading the ways in which heteronormative values and practices are maintained. It then considers the more numerous works addressing alternative sexual politics. Scholars have built on the early critiques of the heteronormativity of social relations and spaces to establish nuanced readings of how powerful discourses and practices continue to maintain hegemonic (even if problematic) notions of heterosexuality. Binnie and Valentine (1999) have shown that little geography has been written to date on the hegemony of heterosexuality (either in society or in the discipline of geography). More critically, Grant (2000: 76) calls for an end to the 'ongoing non-examination of compulsory heterosexuality in geography'. Philip Hubbard's (2000) work addresses these concerns and provides a watershed in geography of sexuality as he argues for a more widespread interrogation of heterosexuality. Hubbard (1998, 2000) notes the ways heterosexuality is normalized through performances, territorial control and moral constructions. Just as masculinity and whiteness have been made visible and the subject of explicit critiques in geographies of gender and ethnicity, so too Hubbard's work begins to map out the discursive and material ways spaces (especially cities) support and encourage certain forms of heterosexuality. He notes that particular, acceptable forms are supported while other sexual relations are regulated and controlled as immoral, abject and spatially marginal or ostracized. Re-engaging with geographies of sex work, Hubbard (2002) shows how sexual identities are constantly involved in processes of negotiation and self-formation. He also identifies how moral and legal discourses and spatial boundaries intersect with ideas of fear, vice and repugnance to construct spatialized entities such as 'streets of shame' (Hubbard, 2000: 204). More broadly, turning to social, commercial and state structures and discourses that shape work and home environments, Hubbard (2000: 208) argues that morally partial, heterosexual landscapes are maintained for 'playing happy families' based on monogamous, procreating, male-dominated heterosexuality. He states that 'research by geographers has begun to interrogate how the *domestication* of heterosexuality continues to be encouraged by a range of state policies from housing allocation and social security to income tax and family law'.

This work highlights and complements geographies of citizenship where political identity has been shown to construct uneven sexual citizenship between sexualities and between countries (Bell, 1995; Binnie, 1997). Binnie (1997) notes that geographies of migration demonstrate the unevenness of citizenship between countries. Reviewing the European Union in the mid-1990s, he records how countries, such as the United Kingdom, have (until recently) been unsympathetic and discriminatory towards homosexuals, while other states, like Denmark and the Netherlands, have led 'to general improvements in the rights of Europe's lesbian and gay citizens' (1997: 245). Even within this possibility, however, Binnie concluded that heterosexual norms prevail since:

> Reforming the law to include long-term same-sex relationships would ... also reinforce discrimination by favouring long-term same-sex relation-ships (that is, those relationships which are closest to heterosexual marriage). It would also continue to discriminate against single nonnational sexual dissidents who lack a domestic partner. In this respect we should not lose sight of who is included and excluded when we articulate the politics of sexual citizenship. (Binnie, 1997: 246)

Beyond these important new attempts to map and critique heterosexu-ality, alternative sexual politics have been the focus of many recent geographies. This has included attention to power, resistance and queer politics that can challenge and subvert dominant orders. Consequently, a number of geographers have concentrated on the identity politics of gay and lesbian actions. In some cases this has included building more nuanced accounts of gay and lesbian areas in the city, where gay neigh-bourhoods are read for the discursive and material alternatives they provide in contrast to wider heterosexual environments. Costello and Hodge (1999) provide examples of Australian neighbourhoods and sites in Melbourne and Sydney where gay identities, lifestyles and activities are noted in both temporary and longer-term associations. In a complementary way, Lo and Healy (2000) and Rothenberg (1995) have highlighted the spa-tial politics of lifestyles. They have presented contrasting cases in which different lesbians arrange their lives in certain parts of Vancouver and New York to access 'a recognisable social space' (Rothenberg, 1995: 180) carved out from the wider straight population and expectations. Making residen-tial decisions is thus a political act. It provides both economic and cultural resources for these women, who can counter the wider heterosexuality of the city by gaining 'the sense of neighbourhood, the visible lesbian com-munity and the alternative lifestyles that seem welcomed and vital to the neighbourhood' (Lo and Healy, 2000: 39). In the case of Park Slope, the politics of the informal creation of a lesbian social space has included the formation of organized social and sports groups and services supporting lesbians as well as the politically significant establishment of the Lesbian Herstory Archives. In the case of Dupont Circle (Washington, DC), Myslik (1996) has recorded the importance of this urban area as a queer alterna-tive to heterosexist power that is so often expressed through homophobic violence. While noting that harassment can still occur even within this area, Myslik concludes:

> As sites of resistance to the oppression of a heterosexist and homophobic society, ... queer spaces create the strong sense of empowerment that allows men to look past the dangers of being gay in the city and to feel safe and at home. ... For gay men, coping with the presence of violence is an act of negotiating power in society. (Myslik, 1996: 169)

More formally, Forest's (1995) review of the incorporation of West Hollywood as a local political unit is another case of homosexuals 'negotiating power in society'. In this case, identity politics promoting aestheticism, maturity, responsibility, and so forth were mobilized in constructing both a positive gay politics and a place identity for West Hollywood (where alternative residential, economic and lifestyle processes could be encouraged). This case is discussed further in Chapter 7.

Beyond the readings of the spatial politics surrounding residential and community developments, contemporary studies are most recognizable for the queer geographies of direct activism. Focusing on events such as ACT Up, Queer Nation and Gay Pride marches, geographers have used queer and poststructural theory to re-present the resistance and subversion of dominant sexual norms that regularly occur during these activities. North American accounts of this kind are most numerous, and Tim Davis (1995) explains how each of these movements is spatially strategic. For instance, in the case of Queer Nation 'kiss-ins', mock weddings and queer shopping outings, the politics target the usual spatial control enjoyed by heterosexism. They interrupt and undermine the implicit heterosexist assumptions about how spaces (especially public ones) can be used and valued. In a similar way pride marches, analysed in the USA, Australia and New Zealand, have been shown as dramatic political vehicles that spotlight heterosexuality as the 'publicly' acceptable identity and the expectations that homosexual or any other activities be private, invisible and unseen. These events have involved the claiming and using of key urban spaces (streets, plazas and other open spaces) where non-heterosexual sexualities are celebrated and supported in public and spectacular events. Critical readings of these events have been particularly important for:

- recording the transitory success and hegemonic heterosexual context of the parades and marches (Brickell, 2000);
- demonstrating the politics of space as events are stages in different spaces within a city (Johnston, 1997, 2002; Knopp, 1998);
- critiquing the mixed implications of having events removed from the most central (and overtly challenging) sites of a city (Johnston, 1997);
- highlighting the commodification and consumption of sexual identities in the 'managed success' of these events, which still maintain dominant, capitalist, middle-class, heterosexual, masculine interests (Knopp, 1998).

The economic and political critique of gay pride marches shows the transitory and qualified success of these queer challenges. While definite benefits are recorded by geographers reporting on activists' and participants' sense of support and affirmation, the privilege and control of wider heteronormative interests continue to be a key contextual factor in the 'management' and response to such queer alternatives.

6.5 CONCLUSIONS AND CONNECTIONS

While sexuality has been the newest of the four core axes of social difference that have structured many social geographies over the last three decades, it has produced a complex and rich field of work. This chapter has shown that geographies of sexuality match those of other social differences in terms of the breadth of approaches that scholars have employed. The biography of Larry Knopp (presented in Part V) is a good example of this diversity and the dynamic way geographers have theorized and investigated matters of sexuality. While some theoretical approaches have been frequently adopted over time in many forms in human geography, perspectives used in geographies of sexuality mirror those of gender. Where gender geographies drew extensively on feminist theory, geographies of sexuality have adopted wider queer theory to interrogate both the composition and articulation of heterosexuality as well as the variety of alternative ways of theorizing sexuality and sexualized lives.

As shown in the opening of this chapter, students, along with academics, are able to recognize the range of ways sexuality permeates our lives and experiences of space and society, including universities – as Knopp (1999) has recently shown (see Part V). This field of social geography is likely to continue to expand as a number of connections are further strengthened, namely:

• *Multiple difference continued*. Continuing attention to difference has enabled a necessary reflection on the ethnocentric and particularly Anglocentric contexts and foci of most geographies of sexuality to date (Binnie and Valentine, 1999; Hubbard, 2000). This continues to be a major challenge and opportunity for current geographies of sexuality and is akin to the critiques levelled at early geographies of gender and ethnicity (where the foci on white, middle-class women's geographies were challenged and the hegemonic position of whiteness was initially left un-addressed). Some exceptions are provided by Elder (1995) and Skelton (1995) while more recent work is raising awareness of groups and settings that have been invisible in the past, e.g. young people's sexuality (Ford et al., 1999; Holloway et al., 2000).

• *Closer interrogation of the hegemonic*. Along with geographies of gender and ethnicity, contemporary interrogations of hegemonic conditions, meanings and relations have seen a fruitful and critical engagement with questions of heterosexuality. Authors working from contrasting settings are documenting the institutional and discursive ways in which heterosexual and heteronormative framing of society and space occurs (see Costello and Hodge, 1999; Hubbard, 2000; Little, 2002a).

- *Contemporary critiques of space.* Recent authors have shown the need to critique further the practice and effects of power in space (Knopp and Brown, 2002) and the opportunity to destabilize traditional approaches to space (especially urban space) which privileged territory and visible identities as relatively stable and usually heterosexual (Podmore, 2001). Attention to metaphoric and symbolic space has become as important in some cases as any specific material location (e.g. geographies of the closet – see Brown, 2000). Additional work on the role of cyber-spaces is also stimulating new conceptions of space and sexual relations (Wincapaw, 2000).

- *Connections between space and identity.* Recent geographies of sexuality have included complex and multifaceted conceptualizations of space and the material and culturally symbolic settings in which sexuality is nego- tiated. Intersecting with studies of identity, geographers have highlighted the different spaces in which sexuality is understood, performed and sometimes strategically assembled (Bell, 2000; Elwood, 2000; Hubbard, 2001; Podmore, 2001). This occurs from micro-spaces of the body and individual bedrooms or homes, through to work and commercial spaces, and the broader so-called 'open' or 'public' spaces where myths of equal- ity and inclusiveness are unravelled (Brown, 2000; Johnston, 1998). The processes by which differences (like sexuality) are assembled and performed as identities are taken up in the next chapter.

SUMMARY

- Geographies of sexuality concentrate on the ways sexual desire, behaviour and identities are experienced and negotiated through space and within specific places. These geographies have varied according to different approaches – both those theoretical perspectives broadly affecting all human geography and those stemming from the development of specialized queer theory.
- Early work sought to map and analyse spatial patterns of sexual behaviour. This focused on 'problem' topics of prostitution.
- Radical (feminist and Marxist) approaches presented major theoretical explanations for the unequal power relations

that privilege heterosexuality through patriarchy and capitalism.

- Feminist approaches challenged the normalized but unequal conditions experienced by men and women in the heterosexual expectations that dominates the arrangement of most western life (e.g. paid employment, domestic work, housing access, urban space). And the invisibility of lesbian lives began to be addressed and reversed.
- Marxist approaches highlighted the position and functions of sexuality within capitalist societies and the implications for urban form and processes.
- Recent geographies of sexuality have drawn on poststructural and queer theory to challenge dominant (heterosexual) explanations and highlight the constructed, fluid, unstable nature of sexual identities.
- Recent geographies of sexuality have also highlighted how sexuality can be constructed or resisted through strategies and actions that implicate various spaces and processes (from bodies to national citizenship).

Suggested reading

As suggested in Chapter 5, Jackson's (1992) *Maps of Meaning* is an excellent place to start reading about the formation and control of categories of social difference like race and sexuality. His fifth chapter, 'Gender and Sexuality', includes an accessible account of how sexuality has been approached in past geographic research. More graphic and personal accounts of how individuals must negotiate sexuality in their everyday lives are provided in Valentine's (1993a, 1993b) articles (underpinning Table 6.2). For detailed attention to theory, Jagose's *Queer Theory* (1996) provides one comprehensive outline of the themes and motivations associated with this alternative approach to sexuality. Also from a queer perspective, *Mapping Desire* (Bell and Valentine, 1995a) contains a rich and readable collection of chapters that demonstrate the striking diversity of non-heterosexual life, spatial relations and social struggles. Finally, the recent scrutiny of heterosexuality is an important area to follow up. Hubbard's (2000) article is a primarily theoretical piece but is essential reading for anyone wishing to advance a critique of the wider social and political issues associated with the dominance of certain heterosexual norms.

ACROSS AND BEYOND SOCIAL DIFFERENCE

PART III

While the chapters in Part II of this book have presented a record of core work being completed in contemporary social geography, scholars have wished not only to highlight these differences, but also to look forward. Considering these differences, they have asked why, how and so what? They have also asked where next, and what changes are possible? These questions have encouraged geographers to respect – but look beyond – the specialist research details and conceptual debates surrounding specific differences like class, gender, ethnicity and sexuality. While not discarding them, many social geographers have turned to consider how social differences can be further understood (even challenged and reshaped). As a result, I have structured the third part of this book as a way to investigate the concepts and work that scholars are considering both *across* and *beyond* difference. The following chapters enable us to consider how people and groups may connect and mobilize around particular differences, experiences or goals, using constructions of difference and space strategically in the process.

Chapter 7 commences with identity, a topic that can be both a personal and an accessible way to interrogate how social differences coincide and an accumulate in layers of uneven meanings and relations for individual lives, specific places or even imagined national identities. Chapter 8 focuses on the concept of power as a notion that underlies the formation and struggle over both social differences and expressions of identity. The chapter reviews a variety of ways power is conceived and then highlights both the institutional and informal relations and struggles that social geographers have investigated in their interest with the processes and outcomes of power. Finally, Chapter 9 extends the general interest in power to a specific focus on one outcome of power relations, namely the exercise of social action. The chapter records geographers' approaches to the notion of social action and the investigation of how difference, identity and power are played out when people take action in different settings and forms (e.g. individual acts, formal organizations or more fluid social movements).

Together, these three chapters provide avenues for both recognizing and moving beyond the potential divisions that can occur when we highlight categories of social difference. Moreover, unlike some recent geographies that focus more on spaces and scales (Valentine, 2000) or processes, flows and problems (e.g. consumption, leisure money, crime, poverty or globalization:

Hubbard et al., 2002; Pain et al., 2001), this set of chapters concentrates on strategic concepts that can be used to combine study and imaginings of alternative futures for the social relations, struggles and opportunities that shape our place-specific, spatially-potent social lives.

7 Identity

7.1 INTRODUCTION: RECOGNIZING SOCIAL DIFFERENCE THROUGH EXPRESSIONS OF IDENTITY

How would you describe yourself?

Would this description vary depending on whether you were talking to a friend, grandparent or employer?

Would this description be associated with particular places or types of space?

Would you use a different description in your parents' home, at university, at a police station, or on a holiday in Amsterdam or Bangkok?

Take a few moments to answer these questions in some oral, written or visual way. If you do, you will likely find that you use a variety of descriptors that might convey your class, gender, ethnicity and sexuality, but a whole array of other attributes, including personal interests and qualities together with selections from a wider set of social values and norms.

This type of exercise quickly shows how we constantly negotiate a variety of social differences in our daily lives; and how, across the differences, we experience moments of recognition and identification with other people and groups. Geographies of identity can show how we experience identity through interconnecting formations of ethnicity, gender, class, sexuality and so on (Dwyer, 1999). These geographies also show how processes of identification involve spatial relations, or how *where* we are (in physical, social and discursive senses) will affect the identities we might choose to convey, as well as the way others may identify us.

A descriptive activity like that suggested above can indicate the varied way we live and perform a selection of identities in different circumstances and places. For instance, while initially drafting this chapter, I lived for two weeks in a student college while visiting the geography department at the University of Glasgow. It was the beginning of the autumn term and the energy, activity (and occasional inebriation) of the new students establishing their life in the college meant that I had plenty of occasions to dwell on issues of identity (mine and theirs)! In the week before arriving, I was very conscious of my identity as a mother, for I was about to leave my son and daughter in New Zealand. Once in Glasgow, I see-sawed between this identification as a mother and my professional identity as a visiting academic. I was also conscious of both my shared identity with first-year students as a 'new user' of the college facilities and as someone negotiating access to the university campus. I was equally aware of how I held contrasting identities. I spoke as an Australian who often had to work hard to understand the Scottish accents of these predominantly 'local' first years. I also looked and behaved as a slightly unconventional older woman. A woman who had very short hair, and who tended to walk alone, mainly in daylight, and in utilitarian ways down Byers Rd. This contrasted with the highly social and relaxed way many groups of young male and female students traversed the same street together at all times of the day and night. These experiences reminded me that identities are simultaneously always about both recognition and difference (Hetherington, 1998; Pratt, 1999b). I recognized shared experiences as a resident of the college and as an academic in the department, but I also experienced many differences: from accents, through choices in femininities, to daily patterns of movement. Cumulatively, these experiences reaffirmed for me that identities involve sets of meanings and qualities that we convey (or sometimes set aside) in order to establish recognition and difference with others. They are constantly negotiated and read by ourselves and others, and they change in value and implications in different places.

This chapter outlines how notions of **identity** are used in contemporary social geography and it focuses on the social and spatial ways in which identities are both constructed and contested. The next section (7.2) reviews how identity has been theorized. It explains how recent attention has been given to the constructed nature of identities, having both relational and spatial dimensions. Section 7.3 considers the different ways geographies of identity have been assembled. Note is made of the different scales at which identity has been studied, and the political nature of identities. It shows that many individuals and groups can selectively and explicitly engage with strategic identities – both those they wish to promote and support, and those they wish to challenge or contest. This discussion forms one bridge between the preceding chapters on social difference and these latter chapters, looking at the ways people work with, and across, difference. It draws together the purposeful ways social differences are negotiated and, in looking forward to Chapters 8 and 9, it highlights some of the power relations and actions that are involved when individuals or groups engage with questions of identity.

7.2 CONCEPTUALIZING IDENTITY

Identity has a long history as a concept in philosophy, psychology and more recently social and cultural theory. A sketch of five broadly different views is provided in Table 7.1. The first four thinkers draw on orthodox or structuralist thinking, which, in different forms, sees identity in relation to notions of 'self' as a stable and autonomous entity, or as an entity shaped by the context and structures surrounding one's life. In contrast, Foucault's approach emphasizes the process by which individuals and groups are positioned and identified through *discourse*; where a sense of self cannot be fixed or certain, but instead is experienced in different ways, through a variety of different *subject* positions. Within social geography, work in the 1980s and 1990s had included Marxist geographers' attention to 'class consciousness' and feminists geographers' attention to gender identities and racial or class identities that disrupt broad categories of 'woman' and 'man'. However, concepts of identity have received greatest attention since the 1990s as postmodern and poststructural approaches overtook the popularity of the critical and humanist work of the 1970s and 1980s. Consequently, non-essential and poststructural approaches to identity taken by Foucault (1990) and Laclau and Mouffe (1985) have been widely

TABLE 7.1 CONTRASTING PERSPECTIVES ON IDENTITY

Theorist/Thinker	Perspective on identity
Descartes (Orthodox European Philosophy)	Identity is based on the notion of an autonomous thinking self who can both form independent meaning and act with agency.
Durkheim (Sociology)	As a challenge to individualism, identity for the individual is seen as an outcome of society, including its dominant norms and values, as well as the way its economic life is organized.
Freud (Psychoanalysis)	Identity is based on the complex self that is produced from relationships between the instinctive 'id', the wider cultural and moral consciousness of the 'super-ego' and the individual 'ego'.
Althusser (Structural Marxist Philosophy)	Identity is produced through ideology via social institutions which position subjects within society (e.g. via church, education, media).
Foucault (Poststructural History)	Identity (of self or other phenomena) is constructed through discourses which position subjects (both those being produced through the text and those reading or consuming the discourse).

embraced by geographers (see, for example, Massey, 1999; Natter and Jones, 1997). Poststructural thinking has encouraged investigations of identity as multiply defined and unfixed phenomena, where the self (or others or places) is constructed through discourses and social relations. Identity is thus recognized through both discursive **subject** positions and social encounters.

Universities provide a useful example of **discursive** identity formation. University spaces and promotional materials, regulations and graduation ceremonies involve institutional discourses that can construct a 'student' identity (see Figure 7.1). Alternatively, local and student newspapers are part of a popular discourse that may construct student identities in quite contrasting ways. Each of these discourses highlights certain aspects and differences that are assembled to constitute knowledge of a particular category. Identity involves the meanings and expressions of self (as student), and others (e.g. lecturers, administrators), and places (e.g. offices, lecture theatres, graduation ceremonies) that may appear relatively stable and singular but are likely to be multiple and unstable, and may be challenged or renegotiated in a range of ways.

Identities may be attached to many contrasting phenomena: identity of the self, identification of others, and identification of places (e.g. a town or community identity). Each of these identities is a social construction, developed as social understandings that may define our experiences of difference and sameness, and that will form boundaries. For instance, Wall explains:

> The process of identity is experienced through constructing boundaries between Self and other, the conceptualisation of groupings: an inclusionary sense of shared connection and affinity held by people, contrived in opposition to exclusionary imaginings of difference. (Wall, 2000: 82)

This self/other construction of identity has been a powerful consideration for many social geographers and sociologists, and it encourages a focus on how boundaries work, boundaries that can aim to define self but also create distance from 'others'. Hetherington's account of this process echoes and elaborates Wall's, as he argues:

> Identity is ... associated with processes of self-recognition, belonging and identification with others. Identity is also a way whereby we create forms of distinction between ourselves and those who we see as being like us and those who we see as different. We generally do this by creating divisions between those with whom we identify and those with whom we do not. Identity, therefore, is how we do membership and how we include or exclude others from membership of a particular identification. (Hetherington, 2000: 92)

While sociologists may concentrate on the social processes of such identification, social geographers can combine an interest in both the social

FIGURE 7.1 CONSTRUCTED STUDENT IDENTITIES
Source: University of Otago.

processes and the spatial dimensions involved, for space is implicated in constructing an identity, drawing boundaries or distinguishing/excluding others. Massey explains this by stating:

> [W]e make, and constantly remake, the spaces and places and identities through which we live our lives. This applies to the ways ... we construct our personal and communal identities ... to how we construct the spaces of 'home' and of 'employment' and how we negotiate the power relations and the boundaries which exist between them. (Massey, 1999: 291)

As will be shown in section 7.3, these spatial considerations are a major part of contemporary geographies of identity as different scholars have considered the personal through to national and international spaces that are constructed and negotiated in forming – or performing – identities.

Identity is 'done', therefore, through discourse and through space via dynamic processes of formation and expression. In terms of identity formation, two dimensions have stimulated and shaped geographers' work. First, identities are seen to be formed through discursive processes or narrative. By attending to the narratives individuals and institutions construct, it is possible to document and read (or deconstruct) the 'narrative constitution of identity' (Somers, 1994). Somers argues that narratives enable us to locate ourselves on the basis of certain collections or meaning:

> [P]eople make sense of what has happened and is happening to them by attempting to assemble or in some way to integrate these happenings within one or more narratives. ... [P]eople are guided to act in certain ways, and not others, on the basis of the projections, expectations and memories derived from a multiplicity but ultimately limited repertoire of available social, public and cultural narratives. (Somers, 1994: 614)

Somers (1994) points to different forms of narrative, including ontological narratives or stories by which we make sense of who we are (and therefore what we may choose to do); public narratives expressed through culture and institutions (e.g. 'working-class hero'); and metanarratives or master narratives by which history and society are recounted, (e.g. Enlightenment, Modernization, Nationalism, Globalization). Somers' perspective has been most commonly adopted in the study of people and personal identity. However, other authors have also noted that 'constitutive narrative' processes can involve the production of identities that are attached to places and communities (Bellah et al., 1985; Entrikin, 1991; Shotter, 1993). In section 7.3 we will see some geographies of narrative identity, where both personal and place-based identities are formed through the different social and spatial relations.

The second aspect in conceptualizing identity formation that has influenced geographers' conceptual and empirical work involves the way identity is practised. Attention to this enactment of identity has centred on concepts of **performativity** or **identity politics**. Identities are performed through bodies, language, dress, actions and spaces. This view encourages the study of identity beyond language and discursive processes. It highlights the idea that identities are practised and articulated, thus different expressive arenas have been considered (e.g. music: Wall, 2000; food: Valentine, 1999; media: Jackson et al., 1999; and work practices: McDowell, 2002b). However, theories of performativity highlight the fact that expression of identity is not necessarily free or voluntary (Butler, 1990, 1997; Gregson and Rose, 2000). For instance, Gonzalez and Habel-Pallan (1994: 82) have noted that 'identity is not simply a matter of choice or free will, but is rather a negotiation between what one has to work with, and where one takes it from there'. The act of 'taking up' an identity focuses attention on the dynamism or active expression of identity. This resonates with Butler's accounts of performativity, for identity can be understood not only via discourse but also through the performance of meanings and categories that involve bodies and actions with social processes and historic contexts. And, as shown in Chapters 4–6, these contexts frequently privilege certain social differences as norms (e.g. masculine, white and heterosexual norms).

Alongside the attention given to performance or articulation of identity, the power relations and politics of identity have also stimulated theoretical and empiric thought. As poststructural approaches have highlighted the unfixed and contested nature of identity, power relations have come into focus – especially their role in contextualizing the formation and practice of identity. Massey explains:

> Identities' ... are temporary ... held together by different relations, with varying degrees of longevity, and so forth. What is (or could be) at issue politically is the power relations through which such identities are constituted and those through which they interact with each other and the wider world. (Massey, 1999: 291)

Identity politics is the most common way in which these power relations are interrogated. Identity politics may be formally practised as different activist groups or social movements strategically construct and promote key cultural identities (or challenge those they wish to see changed). Considerable work has been done in investigating these political choices and negotiations. The spatialization of identity politics has been especially important for geographies of identity (Keith and Pile, 1993). Debates around the way that social struggles over identity take up or reshape/reconstruct space have enabled study of the locations, 'space/time', margins, and alternative spaces of identity politics (Bondi, 1993; hooks, 1990; Knopp, 1998; Massey, 1999; Soja and Hooper, 1993). Examples of how these ideas have influenced geographic studies are detailed in section 7.3.2.

Beyond the view of 'identity politics' as a fairly formalized activist struggle, the term can also be recognized in the less organized and more informal ways in which individuals experience and work with (or around) identities. In these cases, identities can be understood through practices of 'mixing' and 'managing'. For example, Hetherington (1998: 26) argues that 'identity involves combination and the mixing of things at hand, and an ordering associated with that process of mixing'. This can be suggestive of a degree of a freedom that many people will not necessarily enjoy, since some forms of sexuality, ethnicity or disability may involve identities that are constrained or excluded in certain settings. Thus Valentine's (1993b) notion of 'managing' identities is probably a more appropriate argument for understanding some identities. She has shown how gay women have needed self-consciously to manage multiple sexual identities as they move between different social spaces, as will be detailed in the following section.

7.3 GEOGRAPHIES OF IDENTITY

Studies of identity have become increasingly common in recent years as social geographers have been influenced by cultural theories and reinvigorated cultural geographies. While previous 'modernist' approaches to identity focused on relatively structured 'lifescripts' that were thought to shape identity formation (Hetherington, 1998: 23), contemporary geographies of identity have drawn primarily on new cultural impulses and poststructural theories. This has allowed researchers to consider the multiple and constructed nature of subject positions generated by competing discourses, e.g. of 'women' or 'healthy bodies' or 'race' or 'sanity', etc. The work of Karen Morin is an example of this trend in contemporary social geographies and a biography of her approach and research is included in Part V while a detailed description of one of her collaborative works in discussed in section 7.3.1.

Social geography has long included analyses (and classifications) of different types of people and places (e.g. of working classes, or migrants, or neighbourhoods, or ghettos). But, the term 'identity' has become an effective shorthand way to signal some of the key discursive and material qualities or meanings that might be associated with, or re-present, these different groups of people or places. Identity is therefore a useful concept for highlighting the competing meanings that might have broad currency and that distil (or selectively summarize) some features over others. It is also a notion that suggests action to the degree that meanings associated with any particular identity will be conveyed (purposefully or not) in a range of ways, e.g. through dress, language, behaviour, symbols. Attention to these practices and texts mean that social geographers are concentrating on how identities are constructed, re-presented and even struggled over (see summary in Figure 7.2).

7.3.1 Scaled accounts of identity: relational and performed identity

Geographers have investigated identity at many scales. In recent years, geographers have increasingly debated and reconfigured the notion of

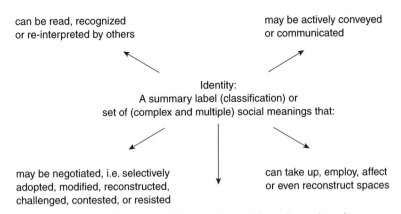

can be read, recognized
or re-interpreted by others

may be actively conveyed
or communicated

Identity:
A summary label (classification) or
set of (complex and multiple) social meanings that:

may be negotiated, i.e. selectively
adopted, modified, reconstructed,
challenged, contested, or resisted

can take up, employ, affect
or even reconstruct spaces

may be represented and performed through a variety of
texts, practices and personal cues (e.g. hair, dress, body language)

FIGURE 7.2 GEOGRAPHIC FOCI ON IDENTITY

scale. For simplicity's sake, I have used the term here to refer to different levels of attention and representation geographers have used in approaching identity. However, scale is *not* seen as *absolute.* Nor are different scales seen to be inevitably or naturally indicative of specific relationships or processes. Rather, scale is recognized as a notion and a set of arenas of activity that are actively *produced* in economic, social and political ways (Delaney and Leitner, 1997; Smith, 1992b). The following discussion notes three of these levels of foci: individual, place and nation. Table 7.2 provides an overview of this type of work. Across these examples we can see the spatial and relational ways identity is formed and performed.

In the case of homeless young people in Newcastle, Australia, Winchester and Costello (1995) reported that the personal and group identities of 'street kid/s' were communicated or performed in social, material and spatial ways. They showed that street kids demonstrated their shared identity through territorial identification (including the coding of spaces with graffiti) and social codes.

The street kids were territorial in their use of derelict land and property and adjacent waterfront both as residential and activity space ... their residential space ... consisted of squats in abandoned warehouses, offices, houses, and other property. ... Their activity space often focused on the SOS centre for meals, and on various squats, corners, and parts of the waterfront as hangouts. ... The group was not only territorial, but had a number of identifiable characteristics and symbolic identifiers. ... In particular, they chose generally to dress in black jeans, T-shirts, and sloppy joes or jackets, wearing Doc Martin boots or Reebok training shoes on

TABLE 7.2   GEOGRAPHIES OF IDENTITY AT DIFFERENT SCALES: SELECTED
EXAMPLES

Scale (Chapter section)	Example (Author/s)	Relational and spatial themes
Individual (7.3.1)	British Muslim young women's personal identities (Dwyer, 1998, 1999)	• Young women experiment with and assemble identities that alternatively (or simultaneously) connect them with family and ethnic communities as well as wider British culture. • Choice of 'Asian' or 'western' dress allows identity to be expressed and actively performed through the spaces of the body.
Individual (7.3.1)	Filipina domestic workers in Vancouver, Canada (Pratt, 1999a, 1999b)	• Domestic workers' identities link gender and ethnicity and class. • Bounding and using space (e.g. bedrooms) can express personal territory and difference.
Individual (7.3.1)	Lesbians' negotiation of identity in Britain and Aotearoa/ New Zealand (Valentine, 1993a, 1993b; Johnston and Valentine, 1995)	• Lesbians express their identity to each other through dress, social interaction and arrangement of personal space. • Lesbians 'manage' their sexual identity and relations within wider heterosexual society and public spaces.
Individual and place (7.3.1)	'Street kids' expression of identity and creation of their own space within Newcastle, Australia (Winchester and Costello, 1995)	• Group identity is expressed through symbols and actions including dress, aggressive behaviour, language, moral codes of mutual aid and graffiti. • Performance of identity is spatial e.g. territorial behaviour in derelict properties and subversion of formal (policed) boundaries and uses of space.
Place (7.3.1, 7.3.2)	Formation of a gay identity for a local neighbourhood in Los Angeles (Forest, 1995)	• An identity (based on a form of gay male lifestyle) is constructed and attached to a specific urban area. • West Hollywood's identity is expressed through meanings and specific spaces associated with gayness (e.g. creativity, aesthetic sensibility and consumption).

(Continued)

TABLE 7.2 *Continued*

Scale (Chapter section)	Example (Author/s)	Relational and spatial themes
Place/ Community (7.3.1)	Expressions of community identity in the face of proposed superquarries in Scotland and Canada (Dalby and Mackenzie, 1997)	• Community identity is formed as a strategic resistance to external mining threats. • Identity is mobilized around symbols (e.g. the Canadian mountain: Kluskap; the Scottish tradition of crafting) to create boundaries between those sharing recognition of the symbol and 'outsider' developers.
National (7.3.2)	Formation of (Finnish) national identity through athletics (Tervo, 2001)	• National identity is formed through narratives of competition and victory in Olympic sports. • Expressions of this sport-based identity are selectively masculine and confine women's position to conservative and traditional identities and roles.
National (7.3.2)	Formation of (New Zealand) national identity through mountaineering (Morin et al., 2001)	• National identity is formed through mountaineering narratives that are gender, ethnicity and class specific. • Expressions of this heroic masculine identity marginalized the many other social groups and cultural values that have sur-rounded mountains and the New Zealand population.

their feet. This in no way distinguished them from many other teenagers. Their behaviour and language [however] were generally more extreme than would be expected from school children. Their behaviour was often aggressive, and included stealing, alcohol abuse ... and foul language ... (Winchester and Costello, 1995: 336)

Winchester and Costello (1995: 338) also documented the social and cultural practice of this street kid identity – through the use of strong moral codes, including 'mutual help and support', 'violent action in defence of its members' and 'respect' for prior occupation of squats by other 'owners'. Together, these practices created a recognizable performance and spatialization of identity and lifestyle.

Other examples of performed identities are shown in geographies of lesbian choices. Sexual identities are constructed and communicated through 'subcultural codes contained in dress, body language and conversation'

(Valentine, 1993b: 244). These identities may be constructed through individual actions and choices while also being read by others who may interpret and recognize them. They are not simply selected by individuals, but rather, lesbians often move between and perform different sexual identities over time and space. Valentine's (1993a, 1993b) work highlights how a variety of everyday spaces create identity conflicts that can sometimes be anticipated and that need to be managed.

> [B]ecause of the taken-for-granted assumption that everyone is heterosexual and the inaccurate images of lesbianism, gay women are often able to 'pass' as heterosexual, and choose when and where to disclose their sexual identity. ... Some women are careful to express their lesbian identity only in public places at specific times. For example several women interviewed feel able to walk through the town centre at night holding hands with a partner because they anticipate they are unlikely to meet anyone they know, but they would not do this on a Saturday morning when family, friends and colleagues are likely to be out shopping. (Valentine, 1993b: 241, 244)

Valentine's work shows that for lesbians identity involves constant selection and management of sexual identities through different social contexts and spaces. And in Chapter 6 (section 6.4.1) we saw how both British and New Zealand lesbians actively managed their dress, the arrangement of their homes and their interaction with friends, family and colleagues in order to express or conceal their sexual identity (Johnston and Valentine, 1995). In this way identities both involve social relations and the negotiation of different spaces.

This combination of individual and external recognition of identity is also explored in the work of Claire Dwyer (1998, 1999). Her research shows how young British Muslim women explore a variety of different identities that are based on notions of femininity, Islam, and Muslim and western cultures. 'Gender, class, ethnicity, racism and religion' are entwined as young women explore and select ways to express a sense of themselves. One way identity was performed involved young women's choices of dress. Dwyer (1999: 20) argues that for Muslim young women 'dress functions as a contested signifier of identity'. Both their own dress choice, and other people's reading of that dress, are full of meanings, depending on whether they were wearing traditional Asian or western clothes:

> 'Asian clothes' are signifiers for young women's religious and ethnic integrity. They draw on representations of feminine (hetero)sexuality, where 'English clothes' signify rebelliousness and active sexuality and threaten religious or ethnic 'purity'. (Dwyer, 1999: 11)

Dwyer's work also highlights the *spatial* nature of identities since the social constructions and choices that are managed by these Muslim women are constantly shifting and negotiated through space. For instance, some young women explained:

> Sameera: I'm constantly thinking about what people will think of me, they must think that I'm really typical. Even when I haven't got a scarf on my head, but I'm like in Asian clothes, I'm so paranoid. Oh people must think typical ... you know that I'm from the dark ages and that. ...
>
> Rozina: If you just walk down the streets and you've got trousers on and one lady says 'I saw her' and that's all they do they gossip.
>
> Shamin: They say 'I saw so-and-so's daughter and she's started going out with boys' ... just because you're wearing English clothes. (Dwyer, 1998: 56–57)

Dwyer's work shows that everyday spaces will need constant negotiation. She illustrates how identities use or take up space, and how they can be differently interpreted in various spaces. Whether they are in their home, in a local street or at school, these young women will be making choices and taking opportunities to maintain, challenge or mix identities. For instance, Dwyer's analysis of interaction in the female school toilets and a school fashion show indicated that these spaces enabled young women to explore the possibility of alternative or hybrid identities – trying things out. These were important experimental spaces that contrasted with the more challenging conditions and expectations of women that were experienced in public spaces such as their local streets or the universities they planned to attend.

Beyond the scale of individual people's lives and personal identities, we can also see place-based and national identity formation. These are very much *social* projects, even though they involve attaching meanings and identity to particular forms of space. In the case of Forest's (1995) study of sexuality and place in Los Angeles, we see how a strategic gay identity was constructed during the lobby for independent local authority incorporation of the area known as West Hollywood. As noted in section 7.2, identity formation can mobilize and project a commonly recognized identity that belies difference, and this was the case where a specific gay male identity was promoted ahead of the diversity of sexual, gender and class differences existing in the population. Drawing on ideas of the narrative constitution of identity, Forest documents the way a specific gay identity was constructed through the press where a gay discourse narrated themes that associated stereotypical aspects of gay identity with the local character of West Hollywood as a self-governing municipality:

> The gay press tried to construct a stable identity for gays by creating a necessary connection between the city of West Hollywood and the 'idea' of gays. ... This connection was established through the construction of a 'constitutive narrative' that sought to establish the 'sexual meaning' of West

> Hollywood, along with elements such as occupations, norms of behaviour, clothing styles, political outlooks, and cultural activities. ... As an ideal city, West Hollywood incorporated the attributes of creativity, aesthetic-sensibility, an affinity with entertainment and consumption, progressiveness, responsibility, centrality, and maturity. (Forest, 1995: 151)

These themes formed not only a constitutive narrative but were also attached to specific sites – the Pacific Design Centre and Santa Monica Boulevard – that were promoted as embodying the existence (or potential) for aesthetic sensibility, entertainment, progressiveness and the like. Here, then, both cultural meanings and specific locations are woven together to associate a particular social group with a place identity.

In a parallel example of place-based identity, Dalby and Mackenzie (1997) have shown that Canadian and Scottish communities will mobilize around symbolic identities that 'present a front of homogeneity' even when great social difference exists. These place-based examples of identity illustrate the strategic and political nature of identity formation and expression and thus a detailed discussion is left for section 7.3.2. A similar case exists for Tervo's (2001) analysis of Finnish national identity formation, which is shown to link the performance of their Olympic athletes with the strategic identity politics surrounding Finland's nationalism.

Before turning to these strategic and political dimensions of identity, however, one further example of scaled investigations of identity involves the study of national identities. A number of geographers have now begun to investigate how particular identities (including specific social differences) are constructed in association with ideas of nationhood (Morin et al., 2001; Nash, 1996; Sharp, 1996; Tervo, 2001). In the case of New Zealand identity Morin et al. (2001) have documented the way differences of gender, ethnicity and class intersect and they show how some dominant categories have evolved in the construction of a national identity that is based on the masculine mountaineering hero. Reviewing the performance of Edmund Hillary and the history of mountaineering, they argued:

> Hillary and the archetypal mountaineer that he came to represent ... came to play a significant role in the construction of national identity in New Zealand ... mountaineering allowed men to display their physical strength, endurance and bravery. ... Hillary's climb was heralded as a major sporting achievement and one to which 'all' New Zealanders could aspire. (Morin et al., 2001: 120, 134)

Morin and colleagues critiqued this association of heroic masculinity with national New Zealand identity. They argued that the portrayal of Edmund Hillary's climb to the summit of Mount Everest was based on specific ideas of gender, class and ethnicity (see Box 7.1). Morin et al. employed ideas about identity as they studied the history of mountaineering, the slopes and accommodation sites at Mount Cook, and the contrasting

Box 7.1 Heroic male identity: conquering mountains

Morin et al. (2001) record how affluent white males' ascent of mountains such as Mount Everest and Mount Cook involve meanings and practices that suggest they conquer and 'knock off' the peak. Practices such as urination can be incorporated in such moments and further extend the contrasting cultural values that might be attached to mountains where men from one culture might revere the mountain and its peak while men from another might seek to show imperialist domination. In the debates around the climbing of Mount Everest, Morin et al. cite Hansen (1999: 229) who notes:

> On the summit Tenzing buried an offering to the gods in the snow and thanked the mountain in prayer: 'I am grateful, Chomolungma'. Hillary took photographs, urinated onto the peak and told another climber, 'We knocked the bastard off'.

Source: Judy Horacek 1990, www.horacek.com.au. Reprinted with permission from *Life on the Edge*, Spinifex Press, Melbourne 1992.

Maori values of the same mountain – known and respected as Aorangi or Aoraki. (Throughout the paper, Morin et al. (2001) consciously use the terms Mount Cook and New Zealand to emphasize the 'colonial relations inherent in ... mountaineering' and the lack of recognition of Maori culture and rights.) They also documented the variety of men and women who climbed, guided and worked in other ways on and below the mountain. They concluded that a hegemonic and heroic masculine national identity marginalized other values and work. Maori values of mountains had no place in this identity, and women mountaineers who climbed Mount

Cook and other peaks were seen as a direct threat to the status of the mountain: 'If women could climb it, then the mountain lost its status as a dangerous and powerful space in which New Zealand masculinity could be tested' (Morin et al., 2001: 133). Likewise the guide work, together with the service and reproductive work that maintained ancillary support and accommodation had little place in the narration of a heroic, mountain-conquering national identity.

Throughout this discussion, geographies of identity have highlighted:

- the way identities are formed, expressed and negotiated at personal, place-specific and national scales;
- the narrative ways identities are constructed through discourse;
- the performative ways identities are expressed in practice;
- the spatial ways identities are conveyed or negotiated through spaces such as bodies, homes, workspaces, public sites and physical environments;
- the strategic ways identities can be mobilized for political purposes; and
- the homogenizing ways particular meanings are mobilized and privileged in identities while other social differences and cultural meanings are submerged, ignored or actively marginalized.

The final point above registers the 'uneven power relations and boundaries' that Massey (1999: 291) associates with identities. Geographies of identity show how people explicitly or otherwise negotiate power relations in a variety of settings. If we consider how people choose to construct, perform or challenge these identities, we arrive at the second broad theme in this field, namely, analyses of powerful and strategic identities.

7.3.2 Powerful and strategic identities

The acts of constructing, relating to or performing different identities are not neutral. In choosing how to adopt, challenge or resist different identities we engage in the politics of identity. We are often strategic in our choices and we may even choose to contest identities and their meanings. At its simplest, this politics arises as we choose to represent certain meanings through the identities we relate to. These meanings carry contrasting values and implications for different people and in varying places – thus we are immediately launched into a personal politics of enacting identity. At an individual level, we saw in the last section that people could be strategic in expressing 'street kid', British Muslim and lesbian identities through their choice of dress, behaviour and use of space. Strategy was employed because various unequal power relations were negotiated by these groups, including processes such as law enforcement, control of ethnic and religious norms, and dominance of heterosexual culture and expectations.

More commonly, geographies of identity politics have been studied where collective identities are developed or contested for an explicit reason. We can see this illustrated in several examples listed in Table 7.2, namely: the formation of West Hollywood as a 'gay city' (Forest, 1995); the mobilization of community identity against super quarries in Scotland and Canada (Dalby and Mackenzie, 1997); and the construction of a gendered nationalist identity through Finnish sport (Tervo, 2001).

In the place-specific case, Forest's (1995) work indicates how a gay identity (that subsumed other identities of class, ethnicity, and so on) could be strategically constructed so that West Hollywood was recognized as a local government entity embracing and promoting affluent gay male lifestyle and culture. Forest (1995) showed that a specific (affluent gay male) population and specific sites and spaces were highlighted as representative of the area. But he notes: 'The highly visible expression of gay men in West Hollywood [is] in contrast to lesbians. ... The symbols and meanings used in the gay press are also almost certainly class specific' (1995: 135). Thus, even while making relevant conclusions about the creation and perpetuation of important links between place and identity, Forest promotes the need for more critical analyses of the uneven power relations surrounding strategic identities.

Analyses of identity and power are furthered by geographers who concentrate on specific examples of power struggles and resistance. Dalby and Mackenzie's (1997) study of community identity as a form of mobilization against threat is a case in point. When super quarries were proposed for Klusap (Cape Breton, Canada) and Roineabhal (Isle of Harris, Scotland) the affected local residents mobilized a sense of community as they developed strategic opposition to the mining. Community identity was purposefully and explicitly assembled as a 'form of resistance' (1997: 102). In the Canadian case, resistance to the mining of granite was couched in terms of a struggle to protect and preserve the spiritual and cultural quality of Klusap Mountain, including a cave that was sacred to First Nation culture and religion. In the Scottish case, resistance to the mining of anorthosite was focused on respect for, and preservation of, Roineabhal and crofting as an historic way of life. Dalby and Mackenzie note that strategic identities were formed that could counter the progress promised in the planned developments. These were identities that drew on a 'deeply embedded sense of place, of local identity, and of spiritual values centred on Klusap and Roineabhal' (1997: 104–5).

As noted in section 7.2, however, identity formation also constructs boundaries between that which is common and that which is different or 'other'. Strategic identities are no different. Indeed, these Scottish and Canadian examples show how differences based on ethnicity, gender, education, politics and occupation all operated under the apparent collective resistance to the quarries. The power of identification through symbols provided one way across these tensions:

> The process of creating community identity frequently relies on mobilisation around a set of symbols. ... Symbols of community identity present a front of homogeneity *vis-à-vis* the other, a 'commonality', a necessary strategic resource in the face of perceived danger. Yet these same symbols are generally sufficiently imprecise or ambiguous that, although on the one hand they form effective 'boundary markers' for the community, differentiating it from the outside and demonstrating its distinctiveness and its common purpose, on the other they provide scope for individual or group difference. (Dalby and Mackenzie, 1997: 102)

This type of geography shows how 'the social' and 'the spatial' share simultaneity (Massey, 1993). Strategic identities that mobilize powerful symbols in the face of larger threats rest on the combination of key sites with culturally and historically significant social values and meanings (sometimes covering major axes of difference). Exploration of these tensions between difference and identity are taken further in Chapters 8 and 9 as a closer consideration is made of how differences (and the identities that convey them) are played out through power relations and social actions.

An alternate form of powerful and strategic identity politics is provided in Tervo's (2001) analysis of Finnish identity. Independence as a nation state (in 1918) corresponded with social and political processes that constructed:

> a national identity and consciousness in Finland. ... National socialisation [was] mediated by numerous administrative, cultural and political institutions. ... The aim of such institutions [was] to spread knowledge of national symbols, narratives, traditions and rituals to people and to fuse local and personal experiences with the national experience, so that territoriality, nationalism and discourses of national identity [became] part of people's everyday lives. (Tervo, 2001: 358, 359)

Tervo's work documents the contribution made by journalism and Olympic sports to this social project of national identity politics. As noted in the New Zealand mountaineering example, discursive and cultural links were made between sporting achievements and heroic nationalism. In a similar vein to the 'heroics' of war, Tervo (2001: 360) argues that 'sports journalism played a powerful, intermediary role in the Finnish national project, as it made the images of sports heroes and victories part of people's everyday lives and thereby gave people new ways in which to identify themselves with their national community'.

As with other geographies of identity, this case illustrates how constructions and boundaries of an identity are also predicated on privileging some axes of social difference and managing or marginalizing others. For the strategic Finnish identity, masculine aggression and duty were dominant, while the position of women Olympians was marginal and where recorded – in connection with gymnastics – it was controlled in a sexist and patriarchal way:

> Sports were on a par with military service in that they were both providers of masculine identity in Finland. To represent, defend and fight for a nation state – whether on the battlefield or in the sports arena – was thus considered exclusively a male duty. ... Gymnastics, in particular, was considered a commendable activity for women, as it maintained and enhanced traditional feminine qualities, such as harmony, beauty and suppleness. ... The aim of physical education was to strengthen the female body in such a way that women could grow into dignified daughters and healthy mothers of the nation. (Tervo, 2001: 365)

Tervo also noted that class struggles and narratives that pitted Finnish values and identities against other nations formed a complex politics of identity. For this discussion, however, the study reinforces the previous themes regarding scale, discursive construction and strategic mobilization of identity. In defining a common identity (for nation, social group or place), narratives and territories are defined while differences, exclusions and silences lie just below the surface.

7.4 CONCLUSIONS

This chapter has presented the notion of identity as one way in which individuals and groups live out – but negotiate across – multiple social differences. While undoubtedly the British Muslim women and the American gay men of Dwyer's and Forest's research are heterogeneous groups affected by diverse material conditions and social differences, a focus on identity shows how meanings of self and others can be assembled, constructed and communicated in homogenizing ways. Originally I noted that such constructions rest upon the identification of our lives in relation to others – the formation of self and other, or of 'we' and 'them' (Pratt, 1999b: 156). Hetherington effectively sums this up by stating: 'It is through identifications with others, identifications that can be multiple, overlapping or fractured, that identity – that sense of self-recognition and belong with others – is achieved' (1998: 24). But recognition and differentiation alone do not create identity. Consequently, this chapter considered how identities are actively constructed and performed through phenomena as diverse as dress, behaviour and media discourse.

Equally importantly, this chapter has demonstrated how recognition and performance of identity involves a range of spatial issues. Examples illustrated how identities are often involved in place-making (e.g. a place for gays: Forest, 1995; or a community to be preserved: Dalby and

Mackenzie, 1997). Other examples illustrated how identities are also often explored and managed through specific types of space (e.g. heterosexualized public space: Valentine, 1993a; or 'westernized' secular and permissive spaces beyond the Muslim home: Dwyer, 1999).

Together, material, discursive and spatial constructions enable people to convey meanings about identities. They also provide opportunities for people to mobilize meaningful identities for political purposes. These strategic politics of identity range from individual expressions of agency (e.g. in clothing choice) through to nationalistic projects. But in all cases, the identities being explored or performed are underpinned with culturally and historically specific power relations, which result in a finite repertoire of choices. Some meanings and values attached to identities are applauded and privileged while others are constrained as much as possible.

Some of these issues of power are taken up in more detail in the following chapter, for power shapes the relations of individual social difference such as class and gender, but also stretches across them to form terrains and practices where privileged meanings and practices are promoted beyond, or over and above, multiple forms of social difference. Thus it is timely to close this chapter acknowledging that identities are never divorced from power, but instead they signal a variety of power relations, constraints, capacities, possible opportunities and politics that require negotiation.

SUMMARY

- Geographies of identity consider how people and places are linked with certain meanings through both social and spatial processes.
- Identities enable recognition of meanings and attributes that are presented or performed through processes as varied as dress, behaviour, use of space and expression of culture.
- Identities also highlight difference since the formation or articulation of identity involves constructing and promoting certain attributes (of self or place) while expressing difference or distance from others.
- Identity formation, in the construction of sameness and difference, can result in boundaries being formed and exclusion being expressed.
- Theoretically, notions of identity have traditionally been linked to ideas of self and subjectivity – either as relatively fixed autonomous entities or as the subject of wider social structures and ideologies.
- More recent poststructural concepts of identity challenge any essential essence or stability associated with self or identity. Instead, they concentrate on the multiple identities we negotiate via contrasting discourses and identity performances or politics.

- Processes of identity formation are linked to the physical, social and discursive contexts surrounding people and places. Discursively, identity formation can be studied as a narrative process (Somers, 1994), as people assemble meanings about themselves, others and places from a repertoire of social and cultural narratives.

- Identity formation can also be understood as a dynamic process of performativity (Butler, 1990, 1997). In this view, identity is not necessarily a free performance of choice but rather an expression of identity/identities that will be historically embedded and mediated through power relations and varying spatial contexts.

- The expression and practice of identity can also be highly political. Identity politics enable the construction and promotion of key identities for specific purposes.

- Geographies of identity have been written at a variety of scales. Detailed studies of bodies and individual or personal identities are common as are those of community/neighbourhood and national identities.

- The formation or struggle over these scaled identities has also shown the social character of these meanings, where social differences are expressed, managed or mobilized for personal and/or collective interests.

- The expression and practice of identity can be explicitly strategic. Geographers have noted how collective and place-based identities mobilize certain constructions of identity for a range of social and political purposes.

- Both the formation and politics of identity is deeply spatial. Geographers have noted how different spaces influence the expression of identities while processes of identity formation and struggle also engage with (and sometimes reconstruct) spaces for political ends.

Suggested reading

One accessible way to start engaging with the concept of identity is to read Kevin Hetherington's (1998) *Expressions of Identity: Space, Performance, Politics*, especially the first chapter – 'Identity Spaces and Identity Politics'. To understand more about how identities can be constructed and communicated through narrative, see Valentine's (2000) article on how school children explore and experiment with identities. In a complementary way, Dwyer's (1999) article outlines her study of young British Muslim women while she argues that identity is performed and negotiated as a dynamic process at a personal scale. Turning to the more political or strategic dimensions of identity, several levels of reading can be attempted. An early theoretical consideration of identity politics is given by Bondi

(1993) in a chapter within a key compilation of works, *Place and the Politics of Identity* (Keith and Pile, 1993). Forest's (1995) article on West Hollywood outlines how a neighbourhood can be associated with a specific (gay) social identity, while Morin et al.'s (2001) and Tervo's (2001) articles outline clear examples of how several layers of social difference can be entwined into specific forms of national identity.

8 Power

8.1 INTRODUCTIONS: UNDERSTANDING SOCIAL DIFFERENCE THROUGH QUESTIONS OF POWER

While the past five chapters have each had a different focus, all of them have shown how social geography engages with issues of social difference. The way societies unevenly structure life based on those differences – be it access to housing, experiences of public space, negotiations of racism or whatever – always involves aspects of **power**. This chapter considers the notion of power as a way to understand the unevenness that occurs between classes, genders, ethnicities and sexualities, and as a means to explain and potentially challenge the way powerful knowledges are constructed and unequal power relations are maintained.

Far from being the exclusive property of the subdiscipline of 'political geography' (Painter 1995), power is one of the broadest and most crucial concepts human geographers of all persuasions must encounter as we build understandings of how and why human interaction and socio-spatial relations occur as they do. Massey (1999: 291, original emphasis) summarizes this in saying: 'always what is at issue is *spatialized social* power: it is the power relations in the construction of the spatiality ... that must be addressed'. Considerations of power in social geography have developed in at least two ways. First, it has been a central concept within individual geographies of difference, when constructions and implications of class or gender or ethnicity or sexuality have been investigated. For example, in Chapter 4 we saw how Whatmore's (1991) discussion of women's unequal experiences in a 'domestic political economy' of the farm were an expression of economic and gendered power relations, while in Chapter 5 we noted Berg and Kearn's (1996) account of the discursive power of 'naming and norming' the racialized and Europeanized landscape of 'New Zealand' via place names. Second, social geographies of power have provided rich analyses of how social formations and problems stretch across, and weave together, diverse categories of social difference. Just as we saw in considering identity (Chapter 7), social geographies of power are synthesizing projects that often tackle the messy, mixed-up and mutually constitutive qualities of social differences through various formations (e.g. group identities, 'communities', 'places'), social relations and struggles (e.g. contests over material conditions, social participation and key sites). Considering how power is understood and experienced enables us to see that these dimensions overlap and influence each other.

This chapter outlines some of the ways social geographies of power have addressed social differences as they are defined and contested in different situations. This work commenced in Chapter 7 as we saw the politics involved in strategic identity formations. It will also continue in Chapter 9 as specific types of social action and different spatial struggles are considered – each implicating expressions and relations of power. For now, however, this chapter takes time to consider theoretical approaches to power (outlined in section 8.2) and then (in section 8.3) it shows how different social geographies document the complexities of what Sharpe et al. (2000) call geographies of 'domination, resistance and entanglements'. To complement this chapter, Part V presents a biography of Michael Woods, a social and political geographer who has worked with various approaches to power. This highlights the different ways he has undertaken power, ranging from geographies of formal politics to that of informal movements. While some of his work is discussed in detail in section 8.3.1, this broader biography provides an illustration of a social geographer who works across a number of concepts and theories, addressing questions of power within a variety of socially relevant situations.

8.2 APPROACHING POWER: CONTRASTING THEORIES

Power can be understood as an ability to achieve goals, results or specific ends. However, the concept has been theorized and debated at great length both within and beyond geography. In the past, social geographers' interests have often drawn on theories that conceptualize power as a capacity or resource that can be possessed or shared or even lost. In part this has been a result of the way different groups (e.g. classes) have felt the effects of power throughout their everyday life (e.g. when capitalist classes have controlled conditions of work, income and housing for labouring classes). Consequently, social geographers have concentrated on how power relations shape people's experiences of social difference (e.g. how some classes have the ability to control the living and working conditions of others, or how some ethnicities can feel powerless in relation to others).

Two contrasting classifications of power are shown in Table 8.1. While each perspective involves many contrasting philosophical and political theories, for the purposes of this book, an overview of the two classifications is sufficient for us to understand the different perspectives geographers have considered (and therefore produced accounts from research in contrasting ways).

TABLE 8.1 NOTIONS OF POWER: GEOGRAPHERS' CLASSIFICATIONS

Allen (1997)	Sharp et al. (2000)
Power as an inscribed capacity	Power through domination
Power as a resource	Power through resistance
Power as technology: strategies, practices and techniques	Power as entanglements

8.2.1 Capacity, medium and technology

Allen (1997) identifies power in terms of capacity, medium and technology/ techniques. Power as capacity concentrates on dimensions of influence or the ability to control or decide on how things are organized. In this way power is 'an inscribed capacity' or property that individuals, groups or institutions can possess. Power is seen as a capacity owned or controlled by an individual (e.g. king) or institution (e.g. government) that can commonly be described as powerful, while other individuals or groups are simultaneously considered to lack power or be powerless. Concentrating on social and economic dimensions of power, theorists of class (Weber and Marx – Chapter 3) both invoke forms of this thinking. They concentrate on the capacity of individuals or institutions (in the case of Weber), or on the inscribed potential for capitalist relations to control and reproduce uneven and unequal conditions (in the case of Marx). In sum, this view of power concentrates on capacity as acquired through uneven social and economic relations.

In contrast, Allen's description of the second perspective on power (as a medium) is one that focuses on the resources or means by which ends can be achieved. Allen (1997: 62) notes that this view 'stresses the "power to" rather than the "power over"'. In this sense, power is related to resources that can be mobilized, a view that is developed in both *structuration theory* (Giddens, 1984) and *resource mobilization theory* (McAdam et al., 1996; McCarthy and Zald, 1973, 1977; Oberschall, 1973). Other theories that consider the means by which power is enabling include theories of networks (e.g. Mann, 1986). These latter approaches have been particularly popular with certain economic geographers, concentrating on the dynamics of industrial and globalizing economies (Amin and Thrift, 1992; Dicken and Thrift, 1992).

Power as a technology, or as a series of techniques and practices, forms the third conception Allen identifies. This view rests on the work of Foucault. In this case, power is conceived more as a flow than a capacity or resource. Consideration is given to the way power is exercised and practised. Rather than seeing power as a property or a means to achieve something, this view of power suggests something more elusive but also more widespread – a phenomenon that 'works on subjects, not over them'

(Allen, 1997: 63). This view concentrates on the impact of power or the relations of power between individuals, groups and organizations. It is not so much a possession (capacity or property) as an effect that circulates through a vast range of cultural practices, systems of meanings and technologies. As a form of poststructural thinking (see Chapter 2), this notion of power is one that focuses on the diverse, interconnected or fragmented; the dynamic or shifting; and the discursive.

8.2.2 Domination, resistance and entanglements

Turning to the second classification of theories of power, produced by Sharp et al. (2000), a number of differences and similarities are apparent. In this case, another three main approaches have been identified, based on how geographers have sought to understand and analyse power in different social settings. These approaches to power reflect the wider theoretical trends in social sciences, as tracked through Chapter 2. First, 'orthodox' theories of power focus on domination, and Sharp et al. (2000) identify these as including liberal and pluralist conceptions of power, through to Marxist and other radical accounts of struggle based on core power relations, for example of class or gender. In contrast, a second, more recent, set of theories focuses on resistance and gives attention to **identity politics** and **resistance** movements, seeking to counter some of the hegemonic tendencies of previous approaches. These latter works often take a 'bottom-up' approach to power, seeing agency in community activism, informal social movements and the capacity for individuals or groups continually to construct moments and spaces for resistance and contestation. I discuss both of these approaches in more detail below.

Sharpe et al. (2000) argue that these first two sets of approaches maintain a binary of domination/resistance. Their third classification therefore highlights postmodern and poststructural approaches that have more explicitly recognized the interconnected forms of domination/resistance. Since the groundbreaking work of Foucault, power is conceived more broadly as a series of entangled weavings where relations and tactics (or technologies in Allen's typology) are constantly dynamic and intersecting. These tactics include relations and strategies of domination, resistance and other less dualistic constructions and performances of power (Foucault, 1990; Sharp et al., 2000). This section outlines these three approaches in a brief fashion. However, further readings for detailed study are strongly encouraged and some suggestions are given at the end of this chapter.

Power as *domination* involves conceptualizing power in terms of control and coercion, where an institution or individual has the capacity to exert force over other people and a variety of spaces. Paddison explains:

a widely held view is that a power relationship exists where A is able to get B to do something that the latter would not otherwise do. ... How A is able to bring about this change depends upon the strategies it adopts ... power is a broad concept and ... it has become virtually inter-changeable with such concepts as force, coercion, persuasion, etc. (Paddison, 1983: 3)

Parallel, to Allen's 'capacity' perspective, this view of power can focus on the power of the institutions to 'control' certain spaces (at least temporarily) and coerce people into following particular rules about movement and access to spaces. In Marxist theory, this view of power (as a particular form of material and social domination) enabled geographies to be written explaining how unequal capitalist power relations maintained capitalist classes in a dominant position over working classes – via both direct economic relations and also cultural and political relations mediated through the state. Capitalist power over labourers was understood in terms of capitalists' economic control of wages, profits and prices, as well as the state and wider social processes that promote economic 'growth' and 'progress' and liberal individualist effort and success over collective conditions.

Radical geographies based on Marxism and feminism were also important for developing an increasing interest in alternative *resistance*-based approaches to power, with special attention being given to 'bottom-up' activism that could produce resistance movements. Notions of resistance provided an alternative way to theorize power. Power was recognized not just in institutions and practices of domination (via the apparatus of the state or the control of territory, etc.), but also in the informal, dynamic 'grass-roots' actions of people who adopt a range of subtle-through-to-violent actions in order to counter established power structures and relations. Together with academics working on questions of nationalism and ethnicity, Marxists and feminists participated in wide-ranging studies of activism and social movements that aimed to resist certain conditions and support the emergence of alternative politics and social conditions.

Sharp et al. (2000) explain how both 'resource mobilization' theories and 'identity-oriented' approaches have also sought to explain how resistance occurred through collective action and special social movements. Identity-oriented approaches intersect with social and cultural geographies' broader interests in identity and consequently geographies of identity politics and social movements have flourished in the 1990s (see books such as Keith and Pile, 1993; Pile and Keith, 1997; Routledge, 1993).

Moving on from these two schools of thought, and drawing on much of the critical reflection within geographies of resistance, an important advance in theorizing about power centres on geographers' consideration of the problems of a dominance/resistance binary. First, resistance has been shown not to form in unique spaces 'outside' power (Keith, 1997; Moore, 1997), but rather as a socio-spatial process that is multiply entwined and visible within webs and sites of power (see detailed discussion of Foucault below). As a product of wider postmodern and poststructural theory, geographers have recently become increasingly attentive to the fact that domination and resistance are not discrete and separate. For instance, Routledge (1997a) has shown how Nepali 'resistance' illustrated a range of powerful – sometimes dominating – practices. Such writings have moved away from homogenizing impulses that construct systems of domination as absolute or movements of resistance as heroic and virtuous. Instead they have begun to highlight how practices and power relations within the associated political entities are often far more complex and blurred. Massey (2000) points out there is nothing necessarily morally better about resistance movements *per se*. Indeed, she cautions against 'romancing the margins' without paying attention to the politics of different resistance movements. The multiple nature of power has thus been recognized whereby systems of domination will include moments and practices of resistance, just as resistance movements will sometimes struggle with – or founder on – relations of dominance and force (Massey, 2000; Sharp et al., 2000).

These recent, more 'messy' and multiple approaches to power have drawn heavily on the work of Michel Foucault. Foucault has conceived power not as an object/resource to be held by some and not others in a hierarchical system of society, but as a circulation or a web or net:

> Power must be analysed as something which circulates, or rather something which only functions in the form of a chain. It is never ... in anybody's hands, never appropriated as a commodity or piece of wealth. Power is employed and exercised through a net-like organization. And not only do individuals circulate between its threads; they are always in the position of simultaneously undergoing and exercising this power. (Foucault, 1980: 98)

The metaphoric definition of power as a web also allowed Foucault to argue that analysis of power should concentrate on localized and specific features and relations: '[Analysis] should be concerned with power at its extremities, in its ultimate destinations, with those points where it becomes capillary, that is, in its more regional and local forms and institutions' (Foucault, 1980: 96).

This type of thinking has attracted enormous interest from geographers who could see subsequent analytical implications whereby geographies of power could 'map' the webs and entanglements of power, and the relations and processes by which power can be seen circulating throughout a given society, neighbourhood, industry or group. Furthermore, in considering Foucault's interest in micro-technologies and extremities, and the specificities of different historical eras and places, geographers refocused their analyses on the particularities of different localities, sites and micro-geographies being performed, for instance, through media texts, buildings and bodies.

Importantly for our discussions of domination/resistance, this view of power brought Foucault to argue later that resistance was not somehow external, or other, to power. Indeed, he stated:

> Where there is power, there is resistance ... resistance is never in a position of exteriority in relation to power ... points of resistance are present everywhere in the power network. ... Just as the network of power relations ends by forming a dense web that passes through apparatuses and institutions, without being exactly localized in them, so too the swarm of points of resistance traverses social stratifications and individual unities. (Foucault, 1990: 95–6)

A further aspect of Foucault's theories of power that has been adopted widely in social and cultural geographies involves his ideas about power, discourse, knowledge and truth. Foucault theorizes that power is exercised through the production of knowledge and truth that occurs through the expression of **discourse**. Approaching knowledge and truth as multiple (sometimes conflicting) sets of ideas that are discursively constructed enables geographers interested in power relations to examine how power is practised through discourse (see Box 8.1) and the positioning of people in different subject positions. If truth is partial and discursively created in power-filled ways, then notions of gender, sexuality, race and so forth can be approached not as fact but as historically and geographically contingent social constructions. This set of ideas has allowed geographers to read the competing discourses that socially and spatially construct meanings about different groups of people in particular places. Such a strategy can be implemented to look at examples such as the particular constructions of women or youth or disabled people as developed through government discourses: policy texts and bureaucratic practices. Equally, discursive analyses can and have been made of how community groups and social movements have actively constructed (or appropriated and reconstructed) discourses that serve their own ends and subvert other more established knowledges, and to 'undermine and expose' power (Foucault, 1990: 101). Examples of these works are provided in the next section.

8.3 GEOGRAPHIES OF POWER

Social geographies of power have drawn on each of the theoretical dimensions outlined in the previous section. The sense of power as a capacity has often been attached to Marxist accounts of the social relations occurring in capitalist societies. For instance, the early work of Fincher (1981, 1991) illustrates a Marxist geography of the state that concentrated on how state relations could maintain/reproduce capitalist class relations. Fincher argued that state relations and capacities were historically and spatially contingent. She therefore explained varying but nevertheless controlling capacities of the state to maintain capitalist social relations of production by the provision and control of housing and social service programmes. In a similar vein, feminist social geographies also initially focused on power relations whereby systems of patriarchy and capitalism were seen to

Box 8.1 Power via the discursive construction of knowledge

Foucault (1980) argues:

[R]elations of power cannot themselves be established, consolidated nor implemented without the production, accumulation, circulation and functioning of a discourse. There can be no possible exercise of power without a certain economy of discourses of truth which operates through and on the basis of this association. We are subjected to the production of truth through power and we cannot exercise power except through the production of truth. (1980: 93)

Source: Judy Horacek 1990, www.horacek.com.au. Reprinted with Permission from *Life on the Edge*, Spinifex Press, Melbourne 1992.

create material and social relations by which men's interests and patterns of life dominated in unequal ways over women (Foord and Gregson, 1986). Both Marxist and feminist perspectives on power created a foundation for later research, focusing on resistance. This developed through the 1980s and 1990s and included the investigations of struggle around labour, gender, ethnic nationalism and state-based resistance (Bookman and Morgen, 1988; Clark, 1989; Clark and Dear, 1984; Smith, G., 1985). It extended into a flourishing set of international literatures in the 1990s (see, for instance, Pile and Keith, 1997; Radcliffe, 1999; Radcliffe and Westwood, 1993; Routledge, 1996, 1997a).

Most common contemporary works, however, draw on the 'entangled' poststructural views of power. The following discussion illustrates this through three different types of work. In each case, the entangled practice and technologies of power are tackled, while showing both the discursive and (social) relational qualities of power-filled situations ranging from formal political arenas through to social protests. In the first two examples, formal institutional examples of power are considered (local government and industry control of workers). Geographies of power are read via the discourses that communicate and circulate different meanings and practices through local government (section 8.3.1) and industrial working conditions (section 8.3.2). But institutional examples are not the only focus for social geographies of power. Instead, accounts of power involved in informal attempts at social change are also possible. This is not a simple analysis of resistance, however, since poststructural approaches see societies as multiply and discursively constituted. Keith (1997: 278) argues that '[t]hrough processes of subject making, institutions, individuals, nation-states and societies alike are revealed as plural rather than singular, the composite products of many subjectivities'. While this view might be thought to have a constraining impact on radical politics and research that is interested in social change, contemporary geographies of power have generally moved from binary analyses of domination/resistance to focus on these multiple and composite conditions. For, if groups and organizations are diversely constructed, then spaces open up for action where one set of discourses and subjectivities can be read against (and challenge) other more established ones. Section 8.3.3 provides an example of this type of work.

8.3.1 Power and local government: a social and cultural analysis in rural England

Michael Woods has been working on various geographies of power throughout the 1990s (see his biography in Part V). Woods' (1997) study of local power structures in rural Somerset (England) uses a traditional subject (local government) for a new social and cultural geography. Looking at the history of local government in England, he finds that rural England is affected by a series of periods, each with its own discourses and relations of power, that unevenly connected local populations across cleavages of class, gender, and so forth. Differences occurred between those that were led and those that made up the local state: 'Membership of governmental bodies remains biased towards certain sections of the local community, whose position is maintained through the manipulation of social and professional networks, and the reproduction of discourses of power and place' (Woods, 1997: 454). Using a Foucauldian approach, Woods argues:

Power cannot simply be derived from control over resources. It also comes from manipulating relations, being able to use others' influence and, most significantly, power is discursively constructed. ... Those who may be considered to have 'power' or 'influence' in a locality, therefore, both owe their position to discourses that legitimate their power, but also to their ability to control the production and circulation of the dominant discourses. (Woods, 1997: 457)

Through a study of the local politics in the twentieth century, Woods shows how different discourses played a part in three different configurations of power and local authority (see Table 8.2). Nineteenth-century ideas of social order influenced an initial period where large landowners and gentry dominated local government. A paternalistic form of local government persisted alongside a growing institutionalization of county and parish councils. Power was maintained partly through a culture that framed this rural society within discourses of estates and stewardship and political discourses that constructed the conservative control of landowners as a 'natural order' led by benevolent 'country gentlemen'.

Between 1917 and 1974, the institutionalization of local government increased and other elites joined the landowning gentry in politics (e.g. 'shopocracy' and bureaucrats). Woods argues that a tension between the importance of agriculture and the desires for both modernization and tradition framed much of Somerset rural life as political power became more fragmented, this time managed through discourses of 'community leadership' and 'public service'. Most recently, Woods points to local politics being increasingly affected by the continuing diversity of local interests (including the emergence of the new middle class – see section 3.3.2), the growth of non-elected bodies and the directions of a strong central government. In this case, contrasting discourses rub against each other, focusing on the countryside alternately as a place of production (historically agriculture) and consumption (new, middle-class rural lifestyles). Power is contested as traditional elite and newer service class interests are negotiated.

While the face of local government has changed, Woods shows the continuing power of traditional elites and the growing influence of the middle classes. This study demonstrates that discourses enable political elites to use powerful sets of meaning and connection to establish and maintain influence.

Discourses of power are instrumental in enabling established elites to retain a greater share of influence than their actual resource base would otherwise suggest. ... Large sections of the population – the rural working class and young women, for instance – are effectively excluded through material disadvantage, non-inclusion in patronage networks and non-compatibility with the 'leadership qualities required by discourses of power'. (Woods, 1997: 470)

TABLE 8.2 DISCOURSE AND POWER IN LOCAL POLITICS OVER TIME
(TWENTIETH-CENTURY SOMERSET, UK)

Period	Local government	Discourses of power and rurality
1888–1918	County councils are formed in 1888 and parish councils are established in 1894 but ongoing paternalism persists to the advantage of elite landowners and gentry.	Power operates through a discourse of 'natural order' via the duties and benevolence of the squirearchy and the arrangement of the countryside through landed estates.
1918–74	Local government becomes more institutionalized with the 1933 Local Government Act and gains further responsibilities after the 1945 Welfare State Act. Responsibility rests increasingly with office-bearers and central state.	Power is expressed through ideas of 'public service' and an apparently 'apolitical rural community' while energies are focused on modernizing the countryside via agricultural change.
1974–	Local government is reorganized (1974) and power is shifted to non-elected bodies (1980).	Power is contested between 'public service' processes/personnel and elected representation from an increasingly diverse population as the countryside now combines those interested in (agricultural) production and (lifestyle and tourism) consumption.

Source: (based on Woods (1997: 471).

Thus local politics is not simply about the official positions and elected representatives on councils; it is deeply social and cultural as power is constructed through meanings and practices that enable the majority of society to be regulated and reproduced. Woods' recent work shows that rural power relations are not exclusively the domains of formal government structures. He argues this by focusing on the increasing numbers of people who are being mobilized in a variety of countryside protests that enable people to explore social movements as alternative forms of political participation and citizenship (see Woods, 2003; Part V; and Figure 8.1).

A. The Countryside March: London, 1998

B. Liberty and Livelihood March: London, 2002

FIGURE 8.1   COUNTRYSIDE PROTESTS: INFORMAL POWER STRUGGLES

Source: M. Woods, private collection.

8.3.2 Power and industry: a reading of space and struggle in the South African compound

Crush (1994) provides an alternative study of power in an institutionalized setting, taking the case of the mining compound in South Africa. Combining attention to discourse with Foucault's conceptualization of the technologies and exercise of power, Crush reads the compound as a socio-spatial entity where the formal power of the mining company is constructed through institutional practices at the same time as informal power relations are embedded in a variety of struggles, negotiations and acts of resistance. Formally, the mining companies' various constructions of compound life provide an example of the material and discursive ways in which space was organized to control mine workers and in so doing constituted 'the spatial exercise of power' (Crush, 1994: 207). Informally, the *mteto* was a set of informal compound rules that shaped everyday life among workers and between management and workers. Both of these dimensions illustrate the 'entanglements' of social and economic relations through a poststructural social geography of power.

The mining compound is a specific industrial structure (see Figure 8.2) that illustrates uneven power relations between companies and workers. It also displays the way space can be constructed as a physical arrangement and a discourse of control. Drawing on Foucault's (1973, 1979) study of historical institutions of incarceration that constructed and controlled 'criminals' and 'the insane', Crush explains:

> The compounds varied considerably in size but all aimed to incarcerate workers for the entire period of a contract and to deny them personal access to the outside world. ... Like other carceral institutions, compounds were spaces within which individuals and groups were 'observed, partitioned, subject to timetables and disciplines', designed not as ostentatious signs of (the mining industry's) wealth and power, but as a form of moral architecture for the 'fabrication of virtue' and hard work (Cohen, 1985, p. 208). Like the Benthamite prison, the compounds were meant to create in the minds of compound residents the illusion that they were being continuously observed (Lyon, 1991; Crush, 1992). (Crush, 1994: 305, 307)

Through these spaces, managers could differentiate between, and control, larger numbers of workers. Power through space enabled the circulation of dominant discourses about authority and control that regulated the bodies, identities, behaviour and choices of workers. As shown in Figure 8.2, workers' bodies were incarcerated within the compound when not involved in shift work. Movement and behaviour was controlled by access to work (via the underground tunnel) and the apparently constant surveillance that was possible from the strategically placed compound management buildings.

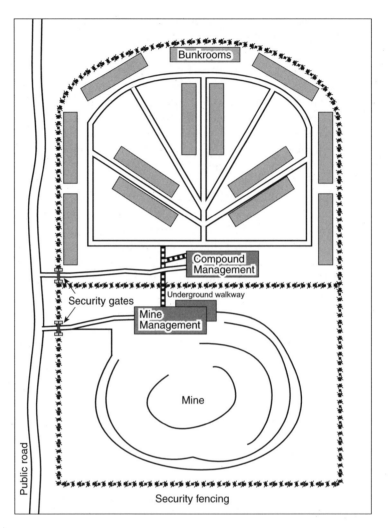

FIGURE 8.2   POWER AND SPACE IN THE MINING COMPOUND

Source: Drawing devised from photographs and analysis published by Crush (1994).

Informally, *mteto* (or compound rules) formed another discourse of power that created a range of socio-spatial arrangements. First, it structured the culture of management control, including both 'coercive treatment' of workers and the limits to which management control could be used before workers would collectively act in protest. Second, and equally importantly, the *mteto* also involved the cultural expression and social interaction that occurred between workers who refashioned their traditional rural backgrounds to establish compound versions of their diverse ethnic identities, beliefs, rituals, songs, stories and leisure activities:

Most miners spent their free time with 'home friends'. ... Workers travelled to the mine with home friends (relatives and associates from the same rural area) and slotted into existing networks when they got there. Home friends relaxed, smoked, drank and played together. ... Although home-friend networks formed the core of drinking groups on and off the mines, they were inclusive rather than exclusive in character. Friendships forged in the ethnically heterogeneous work gangs underground often carried over into drinking and other social activity. (Crush, 1994: 318)

In some cases, well-organized gangs operated within and beyond the mines and these included specific alternative discourses based on criminal interests and 'secret signs and language'. Consequently, much of the social terrain of day-to-day life in the compounds centred on the flows and struggles between different ethnic and gang groups.

Crush concludes that rather than being a linear power relation of industrial control and coercion over a racialized working class, the compound is a good example of the more multifaceted and entangled social character of power. He acknowledges the spatial and discursive authority that mining companies established through the structure of the compound. However, he also points to the alternative and actively resistant possibilities that workers established within this space. He argued:

[C]ompounds are much more than instruments of control or carceral institutions. They are also subaltern spaces: spaces in which black workers carve out hidden domains of social intercourse, experience, and activity which are beyond the powers and comprehension of white compound architects and managers. ... Narratives of struggle view the spatial and cultural form of the compound as a container of domination *and* resistance, of control *and* negotiation, of dehumanization *and* brotherhood, of fragmentations *and* solidarity, of brutality *and* courage. The 'hidden transcript' and lived experience of the compound were far more complex and nuanced than stark propositions of paternalism and control might suggest. (Crush, 1994: 319, original emphasis)

8.3.3 Power and social change: reconstructing the gender of farming

In a different but equally Foucauldian approach, studies of agriculture and gender illustrate how geographies of power can focus on the interweaving of dominant, resistant and other more slippery positions and practices that establish knowledges and political opportunities within an industry and sector of rural society. Work by Mackenzie (1992b, 1994) and I (as Liepins, 1998a, 1998b, 1999) respectively show how Canadian, Australian and New Zealand agriculture is constructed as a discursive and political field involving a number of competing frames of meaning which affect the way food production and women are seen and approached. While each country has dominant scientific, legal and economic discourses that support capitalist and highly industrialized constructions of food production, Mackenzie (1992a), Leckie (1993) and Liepins (1998a) have shown how these industries and sectors of rural society are also explicitly gendered as predominantly masculine.

The masculinization of agriculture stretches across many classes, and differences in ethnicity (and implicitly sexuality). Through narratives of masculinity and femininity, men have been positioned as physically strong, economically powerful and industrially aggressive subjects who contribute to agriculture through visible and valued outdoor work and public, formalized, industrial negotiations (through farmer organizations and commodity and other industry boards). In contrast, women traditionally found themselves positioned in marginal places, sometimes invisibly, as subservient, ancilliary 'helpers' – be this as farmers' wives primarily located indoors, or in minor or auxilliary positions within industry organizations (Liepins, 1998b).

These types of constructions can be studied through a variety of 'micro-technologies' and 'extremities'. However, analyses of rural media are a particularly fruitful source of data. For instance, in Australian and New Zealand media I have found:

> a composite gender identity drawing on constructions of farmers as men who are rugged, physically active in outdoor work, and knowledgeable and decisive in their farm management. The notions of 'tough' and 'active' attributed to this type of masculinity are based on association with hard outdoor labour, control over land and stock, and tenacity in the face of challenging environmental and market conditions. (Liepins, 2000b: 612)

In both Canadian and Australian contexts, some women have actively challenged these positions and have developed alternative political spaces and knowledges in which they can more fully participate in their industries. Foucault's argument that resistance and reverse discourses could be generated from within webs of power – drawing on powerful discourses and inverting them – has been particularly important for Mackenzie's work and my own. Mackenzie has argued that Canadian farm women 'created a discourse of resistance or a reverse/oppositional discourse, which, from the levels of household and rural communities to the stage of provincial and federal politics, dismantled an image of 'farmer's wife' or 'helper' with its assumption of 'free labor' on the farm' (Mackenzie, 1994: 101).

Also using Foucault's attention to the links between discourse and power, I have argued that the 'women in agriculture' movement in Australia has explicitly adopted and at times reconstructed the multiple subject positions and identities of 'mother', 'farmer' and 'business partner' to construct political positions (and space) from which they can speak and participate in industry politics. These identities were often contradictory since significant differences occurred between the women, but as strategic identities they provided political leverage and attention in male-dominated industrial, media and government arenas (Liepins, 1998a; Panelli, 2002a). Using poststructural notions of political subjectivity (outlined in section 9.2.3), I noted how women developed discursive agency and engaged with – and moved between – a number of different legal, agricultural and environmental discourses (Liepins, 1998c). The various groups and texts generated within the movement have enabled women to highlight their multiple identities as 'farmers', 'business partners', 'mothers' and 'carers'. These identities:

> indicated the spaces and rhetorical positions from which women could speak and lobby for the movement. In this way, women continuously moved discursively between their work and concerns in their paddocks, homes, farm offices, and local schools as well as at their agricultural business sites and at the diverse places of their community work and paid employment. (Liepins, 1998a: 141)

Power was realized in these different positions and discursive choices so that women were able to contribute actively to a range of legal and environmental debates, presenting sometimes different, sometimes complementary, perspectives and knowledge to that already circulating in the Australian contexts.

> The key subjectivities of businesswoman/farmer/partner and mother/carer were employed from within broad discourses about farm entrepreneurship and traditional rural femininity. These positions enabled women to speak on environmental issues based on particular constructions of nature, rural space, and womanhood. As businesswomen/farmers in a capitalist agricultural system, women located themselves in a rural space dominated by agriculture as an industry. From this position, and drawing on their experiences as women 'on the land' and in the 'farm office', they articulated a construction of nature based on the biophysical resources they coordinated in their business. In contrast, as 'mothers and carers', women located themselves in more conventional female rural spaces: specifically the 'community' and 'family'. The movement articulated women's perceived maternal and caring qualities such as cooperation, nurturing, education, and communication. (Liepins, 1998c: 1192)

This work shows that power need not only be conceived as something possessed or mobilized by those in dominant positions or through important institutions (e.g. local government or industrial companies). Rather, groups that seek to play a part in society and shape some of the meanings and practices of that society can effectively position themselves and construct knowledge through a range of discourses (see Box 8.2). This can be seen as power in action – and the example of women farmers is taken further on this note in the following chapter.

Box 8.2 The power of 'women in agriculture': shaping gender in farming

Throughout the 1990s, the 'women in agriculture' movement was pivotal in seeking recognition for the contribution women make to Australian agriculture. A key strategy involved the challenge to traditional assumptions about women being 'farmer's wives'. Via conferences, media releases, government lobbying and networking, groups within the 'women in agriculture' movement highlighted women as 'farmers', 'managers', 'partners' and 'business women' as well as 'mothers' and 'carers' (Liepins, 1996, 1998a, 1998c).

One example from the *Australian Farm Journal* (Nicolson, 1994) reflects publicity and features written for the first Women in Agriculture International Conference.

Women are the invisible farmers and their contribution to agriculture is often overlooked by society — something the organisers of the first International Women in Agriculture conference aim to overcome.

'Rural women still face the wider community's perception of a farmer as the stereotyped, bronzed male with elastic-sided boots, a ute and a dog.'

Source: Nicolson (1994: 37).

8.4 CONCLUSIONS

This chapter has shown how notions of power are integral to writing social geographies that can address the way social differences are constructed and negotiated. Different concepts of power have influenced the way social geographies have been written. Section 8.2 showed that structural accounts of power – whether conservative or Marxists – have focused on power as a form of domination or force that frequently resides in particular institutions and the relations these institutions have with groups of people. Where political geography has often concentrated on the nation state and 'its' territory and citizens, social geography (particularly Marxist and feminist geography) has focused on how institutions like the 'state' and the 'family' have enabled the reproduction of capitalist and patriarchal social relations (e.g. Fincher 1981, 1991). These foci on power through dominating relations have supported critical geographies that have analysed social formations stretching across (and linking) difference such as class and gender.

More recently, interests in identity and the politics of resistance, have encouraged some social geographers to analyse how power is practised through social groups and enactments of protest that challenge established meanings and structures. These works focus on power as resistance and include studies of how activists join across some differences (e.g. of class and ethnicity) to construct strategic identities that highlight their politics. These types of work have been important and inspiring for demonstrating that groups who are often positioned in socially marginal ways (e.g. women and homosexuals) should not be seen as 'powerless'. Indeed, margins can become powerful spaces from which to establish counter positions and politics (hooks, 1992; but also see commentary and critique by Monk and Liepins, 2000; Pratt, 1999b; Smith, 1999).

In the last decade, however, geographies which recognize the complex, entangled politics of domination/resistance and centres/margins have highlighted the importance of studying the construction of positions and boundaries, and the multiple power relations which are present simultaneously in instances of domination and resistance (Massey, 2000; Sharp et al., 2000; Smith, 1999). In section 8.3 three examples illustrate how power is engaged and negotiated in different settings, but in ways that highlight the mix of dominant and resistant relations and the entangled and discursive qualities of power. Woods and Crush illustrated how local government and commercial structures can be established as powerful, but in changing and discursive ways that always include the opportunity for challenge, and sometimes subversion. In a different fashion, research into farm women's geographies illustrates how activists develop powerful positions by both engaging with dominant relations and also nurturing resistant impulses. These groups and organizations could not be seen as 'outside' terrains and spaces of agricultural power, but rather demonstrated the potential (and sometime contradictory conflicts) of multiple positions both within and beyond mainstream agricultural arenas.

Geographies of power have come to be depicted as 'entangled' where forces, practices, processes and relations of power illustrate the mutually entangled powers of resistance and domination in 'countless material spaces, places and networks' (Sharp et al., 2000). This more complex view of power also encourages geographies where the spaces and subjectivities of such politics becomes more visible (Keith, 1997: 282) or 'more grounded inquiries into the practices of domination/resistance whereby specific spaces, places or "sites" are created, claimed, defended and used (strategically or tactically)' (Sharp et al., 2000: 28). These types of spatially sensitive commentary can be found in geographies of social action, for it is here that the combinations of complex social differences and the practical choices about political meanings and strategies are displayed in particular activities and places. Thus Chapter 9 provides a link between geographies of power and the material and symbolic implications that follow when people engage with specific social differences and power relations. In these instances, power is rendered temporarily visible and sited within given places and groups that are purposefully constructed.

SUMMARY

- Power is a key notion in many subdisciplines of human geography.
- In social geography, power is central to the process of understanding uneven social difference and the various social and spatial relations involved.
- Considering power also enables social geographers to see the way different forms of social difference intersect in specific groups or places.
- Different geographers have classified the many theories of power, but a useful distinction is made between three sets of thinking:

    - power, conceptualized as a capacity or possession that can be attributed to individuals, positions or regimes, and which encourages an analysis of domination, control and coercion;
    - power, conceptualized as a resource or (potentially) a possibility for resistance where different individuals and groups can mobilize power to achieve their goals and perhaps alter social conditions – both resource mobilization theory and identity politics are used to explain these activities; or
    - power, conceptualized in terms of the technologies, strategies and practices through which power is exercised as a process or a series of entanglements and performances.

- Contemporary social geographies of power frequently take the latter approach and build poststructural accounts that draw on Michel Foucault to study power as it circulates in net- or web-like structures and is articulated through discourses and the creation of various knowledges and practices.
- Beyond theoretical details and debates, social geographies of power can be usefully classified according to the theory adopted, scale of analysis or type of social and spatial foci involved.
- Traditionally, various levels of government have been a subject of both political and social geography. Studies of local politics (such as Woods, 1997) can illustrate how uneven combinations of class and gender can characterize local decision-making and forms of government.
- But social analyses of power relations need not be confined to the state. Instead, the intertwining of formal and informal practices of power can be highlighted in many settings and spaces, including work environments such as Crush's (1994) study of class and ethnicity in South African mining compounds.
- Intersections between power and action can be studied when social movements and social change are considered, and power is seen as actively constructed and discursively contested (as in Panelli's (2002a) account of the women in agriculture movement in Australia).
- Social geographies of power usefully intersect with wider social and cultural geographies of identity and identity politics in order to show how social differences can be strategically assembled and sometimes effectively challenged. This links with broader considerations of social action – the focus of Chapter 9.
- Finally, contemporary social geographies of power continue to highlight the ways diverse spaces, places and networks are implicated and even reshaped through the ongoing interplay of power relations.

Suggested reading

Sharp et al., (2000) introduce their book *Entanglements of Power* with a very accessible introduction and overview to power: the way it has been conceptualized and employed both within and beyond geography. For more detailed theoretical work, reading some original pieces by Foucault is a good idea, especially his 'Two Lectures' (1980) and Chapter 2 (in Part 4) of his *History of Sexuality: Volume 1* (1990). The commentary by Philo (1992) gives one example of a geographer's theoretical engagement with Foucault's work. Alternatively, the article by Crush (1994) used in section 8.3.2 of this chapter is a graphic example of how space, discourse, and social differences are entwined in power relations, using the South African mining example. For those more interested in power as a discursive practice and struggle, many geographers are writing such analyses using different topics and conflicts, e.g. Radcliffe (1999) on state and popular power in Latin America, and Sharp (2000) on politics expressed through the *Readers Digest*. Finally, a good place to start reading about the intersection between power and resistance (a link between this chapter and the next) is provided in the variety of works gathered by Pile and Keith (1997) in their book *Geographies of Resistance*.

9 Social Action

9.1 INTRODUCTION

Have you ever chosen clothes to wear for a strategic reason?
Have you ever carried a knife or other sharp object in case you had to use it?
Have you ever elected not to live in a particular neighbourhood?
Have you ever signed a petition?
Have you ever participated in a protest march?
Have you ever joined a political organization?

Any of the above might constitute a form of social *action* – whether short term or as part of a long-term life choice. As we saw in Chapter 7, some of these decisions and actions can be part of how we articulate our personal sense of self and explore identities. They can also be part of a wider social process that mobilizes people around certain social differences or seeks to stretch across them. For instance, when I first drafted this chapter, my then 15-year-old daughter told me she was going to an anti-Bush, anti-war protest march in support of people in Afghanistan (see Figure 9.1). This was her first action that could be called publicly political and she had her own reasons for wanting to participate. Yet by joining the march, she also became part of a wider temporary group of Dunedin people who wished to join across their considerable differences (in class, age, ethnicity, religion and nationality) to make a common statement. Combining as a protest march (see Figure 9.2), local residents and university students and staff from within and beyond New Zealand, with a broad range of interests and identities, joined together to perform a social action that characterizes many of the definitions, social relations and spatial issues making up the content of this chapter. For in this chapter we arrive at possibly the most challenging and stimulating object of social geography: social action.

Having outlined the many ways social geographers have approached the differences and power relations between people, this book concludes

FIGURE 9.1   GETTING READY FOR A PROTEST MARCH (DUNEDIN, 2001)

Source: Freeman, private collection.

by considering social action as a topic that symbolizes the interrelated and ever-changing nature of socio-spatial life. Building from the previous commentary on social differences (Chapters 3–6) and the way notions of identity and power have influenced accounts of how multiple social differences are experienced and negotiated (Chapters 7 and 8), this chapter turns to analyse geographers' accounts of social and political actions. Geographers' analyses of action highlight how people participate in different processes that shape their identities, lives and societies. These actions can be

FIGURE 9.2 'SAY NO TO WAR'

Source: Otago Daily Times.

understood as political in a variety of ways. They range from formal political actions such as structured protests against government or commercial developments, through issues of industrial participation, to the construction and maintenance of personal identities and lifestyles – and the places in which they can occur. Examples of these actions are discussed at length in this chapter and they show how people can act together to participate within, or challenge and/or change conditions under which they live and participate in their society.

In sum, this chapter functions as both a review of social action studies, as well as a commentary on how attention to social action can support the continued development of social geography. The following section outlines how social action can be defined and characterized in different conceptual and theoretical ways. Then, in section 9.3, a detailed review is provided of the contrasting ways in which geographies of social action have been presented recently. Across a range of examples, generic themes are noted. These include geographers' concentration on social and spatial relations and the processes that are involved when people choose to shape the environments in which they live. These foci enable geographers to discern how people join across (sometimes great) social diversity to produce the types of action and change they wish to see. Finally, the chapter concludes by noting four ways in which geographies of social action will continue to be crucial in stimulating the ongoing effectiveness and relevance of social geography. Part V complements this chapter by presenting a biography of J.K. Gibson-Graham. As introduced in Chapter 3, Katherine Gibson and Julie Graham have explored alternative actions in the writing of geography as well as in the way social and economic conditions can be analysed. Part V provides a more detailed description of these decisions and perspectives on action.

9.2 DEFINING AND UNDERSTANDING SOCIAL ACTION

9.2.1 A definition

Social action can be understood as the acts, practices and strategies people implement individually or collectively to maintain, modify or challenge the structure/s and/or operation/s of the places, settings and societies in which they live. As noted in Figure 9.3, geographers may approach the study of such action by establishing their inquiries at a number of different points, most commonly by analysing the acts and purpose of the social actions themselves (foci (a) and (b)). Nevertheless the contexts and people involved in the actions (foci (c) and (d)) are often equally considered as influential in shaping the stimulus and character of the actions. Finally, it should be noted that the geographies of these actions are always highlighted as spatial issues (foci (e)), such as where the action is taking place and what struggles are occurring over space or the meaning of place. For example, in Chapter 7 the work of Claire Dwyer illustrated how something as 'simple' as choosing clothes to wear is, in reality, an action of considerable significance (see section 7.3.1). She explains:

> the dress choices made by an individual are the result of an intersection of different factors including her ethnic heritage, socio-economic class, parental or familial attitudes, religious beliefs, political affiliations and personal orientations. ... [T]hese decisions are mediated within the specific context of the social (and spatial) divisions at school as individuals negotiate their identities in relation to others. (Dwyer, 1999: 6)

Taking Dwyer's work and using the definitional elements depicted in Figure 9.3, we can observe the following:

- the actions (a) involve deciding on specific dress styles;
- the purpose of the actions (b) consist of young women's wish to experiment with identities and interests in ethnicity, religion and independence;
- the contexts (c) involve a variety of social factors – noted in the quote above;
- the actors (d) as young women are positioned as subjects in many contrasting discourses (e.g. pupil, daughter, Muslim); and
- the spaces involved in the action (e) are integral to the negotiations and consequences each young woman faces (be they the school toilets, the stage of the fashion show, local streets, family homes or 'inside' spaces under the traditional *hijab*).

Definition	Key elements	Foci of various geographic studies
Social action involves the acts, practices and strategies that	ACTIONS	(a)
people implement individually or collectively	ACTORS/AGENTS	(d)
to maintain, modify or challenge the structure/s and/or operation/s	PURPOSE	(b)
of		
the places, settings and societies in which they live	CONTEXTS	(c)
	SPATIALITY (throughout the social action)	(e)

FIGURE 9.3   DEFINING AND STUDYING SOCIAL ACTION

While this is a simple way of identifying a number of aspects of action that geographers study, wider dimensions and theoretical concepts are also involved as the following two sub-sections will show.

9.2.2 The scope of social action

The diversity of social action is immense. Nevertheless we can recognize at least two broad dimensions that characterize the scope of all actions. First, a temporal dimension can distinguish social actions in terms of whether activities are temporary or medium- to long-term endeavours. Second, we can identify the degree to which activities are maintaining or enhancing existing social structures and power relations, or the degree to which are attempting some form of social change. Figure 9.4 depicts these broad qualities and illustrates how different forms of social action might be located across these axes of time and social purpose.

Figure 9.4 depicts three fields of action. The first involves those actions that support the continuing maintenance of existing social relations and structures of power. Melucci's (1996) conceptualization of collective action and the degree to which actions maintain or breach the limits of social systems is relevant here (see Figure 9.5). In the case of the first field of action in Figure 9.4, actions will be organized to operate within, and

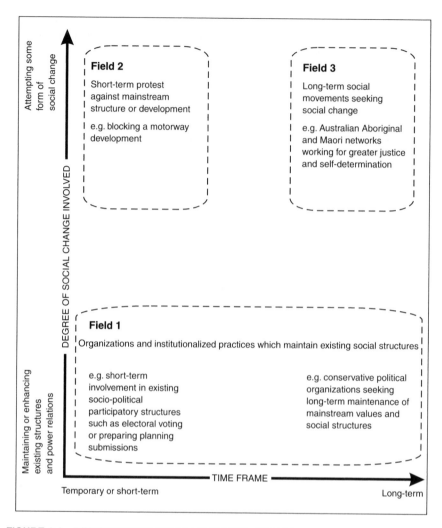

FIGURE 9.4 DEPICTING THE SCOPE OF SOCIAL ACTION

maintain, the structures, relations and limits of a social system involved. Compliance with the legal system of a society and participation in the democratic processes of voting and submissions can be seen as actions by which people participate in maintaining existing social relations and power structures.

Figure 9.4 shows a second field of action which includes those actions that swiftly attempt to challenge or stop existing conditions or developments from within mainstream structures. Single, issue-based protests are a good example of this type of action, where interest and activity flares

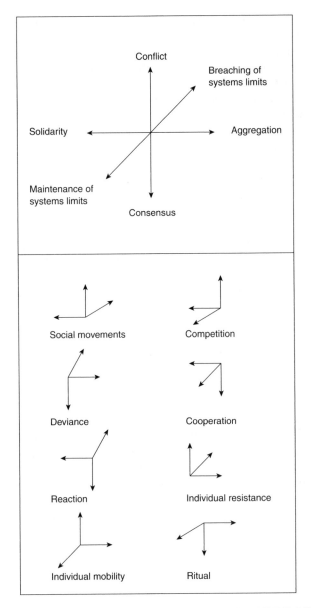

FIGURE 9.5 MELUCCI'S DIFFERENT ORIENTATIONS OF COLLECTIVE ACTION

Source: Melucci (1996: Figures 1 and 2).

quickly in response to a concern and participants work strategically to challenge or halt a situation they oppose. This type of action best illustrates the type of anti-war march presented in Figures 9.1 and 9.2. In 2001, responses to the military action in Afghanistan were widespread but relatively temporary – a longer-term social critique against the 'war on terrorism' is a more complex and challenging action, and better associated with the third field of action. This third field involves individuals, groups and social movements who are committed to extensive social change. Because the agendas are so complex, such actions tend to operate over a far longer period of time. As will be shown in section 9.3, these might include individuals organizing alternative craft economies and social movements associated with socio-political change. In Melucci's (1996: 26–33) terms, whether these changes are classified as 'individual resistance' or collective 'social movement', they will involve social codes being challenged, and the 'breaching of systems limits'. In both cases, participants hold long-term interests in making major changes to the socio-political and/or economic and environmental dimensions of life. They may include individuals in personal negotiations of domestic and economic needs and relations. Or, in terms of collective endeavours, the action may involve larger movements who engage extensively with power relations, including the construction of particular discourses and the use of diverse strategies of resistance or reconstruction. Geographies of each of these actions are discussed in detail in section 9.3.

9.2.3 Approaches and concepts

Variations in the way social action is explained and investigated depends greatly on the different theories and perspectives employed. As shown in Table 9.1, both the people and the actions themselves are conceptualized in a range of different ways. In each case, contrasting philosophical and theoretical perspectives result in different views of the people involved and the contexts and processes by which they establish action.

For instance, Weberian-inspired action theory works primarily from an individualist view of action (Weber, 1978) and categorizes actions according to whether they are traditional (based on past history), affectual (emotions-based), instrumental (goal-based), or value rational (value-goal related). In contrast, Marxist critiques of action are grounded within the broader critical political-economic perspective outlined in Chapters 2 and 3. In this case, people taking action are addressed in association with their class position and described in terms of their class-consciousness (or false-consciousness). Action is then explained in terms of class relations and class conflict where conservative action is seen to maintain class positions to the advantage of capitalist classes, while critical action is seen as class struggle as a means to adjust or address uneven conditions of disadvantaged classes (see section 3.3.2). Marxist approaches are one form of structuralist thinking; Giddens' **structuration theory** is another. Giddens equally views action as affected by structural contexts. However, he also places an emphasis on agency. This enables Giddens (1984) to argue for a 'duality of structure' whereby people produce society (its relations, institutions and events) but

TABLE 9.1 CONCEPTUALIZING SOCIAL ACTION: A SUMMARY

Theoretical approach	Conceptualization of people	Conceptualization of action
Action theory (Weber)	Individuals	Action is classified according to four categories of action: traditional, affectual, goal-rational, and value-rational actions.
Marxist political economy critique	Classes	Action is understood as a function of class relations and class struggle based on modes of production and class consciousness.
Structuration theory (Giddens)	Individual agency	Action is understood to be a result of people using their agency within the structural contexts of their social setting via rules and resources.
Actor network theory	Actors or actants	Action is seen to occur via networks as actors define problems, enlist interest from others and enrol them into a mobilization to implement a desired action.
Poststructuralism	Subjects	Action is constructed as a process by which people draw on the conflict or potential within different discursive subject positions, knowledges and spaces.

are also influenced, even constrained, by it. Consequently, action is a product not only of people's agency, but also of the 'rules and resources' of the particular social system in which they live.

Finally, the two theoretical approaches that have gained widespread adoption in recent social geographies of action involve **actor network theory** and poststructural accounts of **political subjectivity**. Actor network theory draws on sociologies of science. It views action occurring via individual human, composite (e.g. institutional) and non-human actors or 'actants', who identify problems and then enlist others into a network to achieve an intended outcome (Callon, 1986; Law, 1994). This enrolment is known as a translation process and involves actors being attracted to a project or problem, enrolled and mobilized to take action. Geographies based on this approach are discussed in further detail in section 9.3 (see especially the work of Murdoch and Marsden, 1995).

Poststructural perspectives form the last influential approach to geographies of social action. Since poststructuralists do not see the human subject as a discrete and stable individual actor, they concentrate on the subject positions constructed through discourse and the articulation of power relations. The diversity of subject positions set up contrasts, conflicts and choices for people to work through and select, as Davies explains:

> The speaking/writing subject can move within and between discourses, can see precisely how they subject her, can use the terms of one discourse to counteract, modify, refuse and go beyond the other both in terms of her own experienced subjectivity and in the way in which she chooses to speak in relation to the subjectivities of others. (Davies, 1991: 46)

A poststructural view of political subjectivity concentrates on how certain positions are strategically highlighted and promoted and others are challenged or put to the background. Liepins' commentary on 'women in agriculture' (in Chapter 8) provided one example of this work. In the following section, C. Gibson's (1998) account of aboriginal musicians also draws on a poststructural discursive view of action to show how popular music can form a complex discursive field in which an alternative indigenous geopolitics can develop. Likewise, Gibson-Graham's research (1995, 1996a and 2003) – detailed in Part V – illustrates that poststructural views of political subjectivity can be used to recognize how different people respond to economic conditions and the possibility of rearranging work and living arrangements beyond dominant capitalist wage-labour ones.

9.3 COMPARING GEOGRAPHIES OF SOCIAL ACTION

Geographers have long sought to engage critically and analytically with the social concerns and relations that result in people taking action over various matters. Early work in what might be called modern social geography includes analyses of social protest against foreign military policy. For example, in a sobering echo from past concerns, current critiques of 'war on terrorism' can look back to the 'Confront the Warmakers' protest and Pentagon March of October 1967. Akatiff's (1974: 26) account of this earlier march illustrates the systematic *spatial science* culture in which he wrote his geography, arguing that 'geographic analysis of distribution, dispersal, flows and environmental perception provide powerful analytic tools for the general understanding of these movements ... for social change'. Even though a 'spatial science' flavour to geographies of social action is not common any longer, he nevertheless foretells of the entwining of social and spatial dimensions that have continued to guide many geographies of social action. Indeed, while contemporary geographers employ many different approaches when theorizing or analysing empirical data on social action at the turn of the twenty-first century, two generic qualities emerge. These involve, first, the attention given to social processes by which people move from experience and difference through to an engagement with identity, power and action, and, second, the attention given to the socio-spatial qualities and relations of the social action – both in terms of the settings and also the ways spaces and places are strategically shaped or consequently influenced by actions.

9.3.1 Social processes: from difference to action

In considering social action, especially that devoted to some form of change, geographers are faced with the need to recognize how people's diverse circumstances and experiences nevertheless result in sufficient stimulus for specific purposes and identities to form, through which power relations are engaged and actions are chosen and implemented. Thus, to act – to participate, contribute, challenge or resist – involves intersecting social processes where difference, identity and power are negotiated.

People (as actors, subjects, agents, etc.) experience issues that concern them and come to recognize differences and commonalities that provide them with opportunities. They can consider their own, and others', different positions/roles/subjectivities. Equally, they may become aware of the existing power relations/networks/discourses surrounding them and may choose to mobilize their positions and concerns through engaging in different ways with these power relations. One example of these processes is shown in the work of Murdoch and Marsden (1995), who trace the proposed development and opposition of expanding mining operations in Buckinghamshire, England. This is an example of social action within the first field of action (Figure 9.4), where the development – and responses to it – occurs within the existing structures and relations of a social system (as opposed to challenging it).

Analysing this local mining and environmental conflict, Murdoch and Marsden employ **actor network theory** to identify the different groups of people and their interests and actions. The local planners and national mining company personnel are conceptualized as two sets of **actors** who, for different planning and commercial reasons, wish to support the investigation and development of mineral extraction in Buckinghamshire. Participating within the existing economic and political structures and networks, these parties mobilized the existing quota system and county interests to spread extraction activities more widely from the previous concentrations in the south of the region. They respectively proposed and supported new developments of mineral extraction in a relatively untouched northern part of the county (see Figure 9.6). In opposition to these initiatives, local residents of the affected area formed a unity of shared concern for their area. A 'translation' process occurred as they connected with – and 'enrolled' – upper-class, political and cultural networks at national levels in order to oppose the developments. Working within the planning appeal process and the national media, and drawing upon other non-local actors and networks (heritage, gardens and private school old

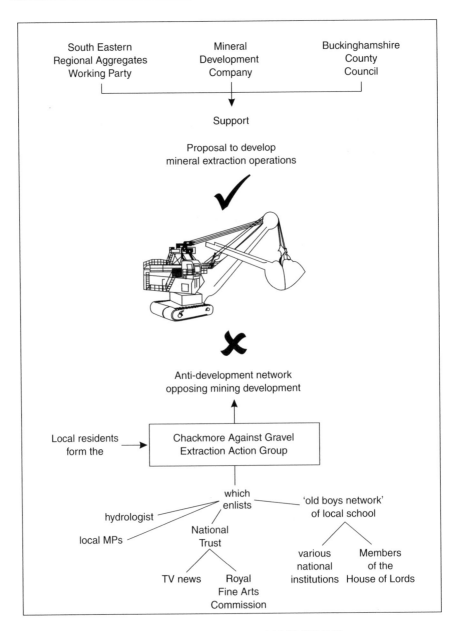

FIGURE 9.6   NETWORKS OF ACTION FOR AND AGAINST MINING

boys networks), the local action group was able to counter and halt the investigation. In this case, Murdoch and Marsden argue that:

[t]he outcome [i.e. successful opposition] depended on which set of associations remained strongest and which meanings [i.e. economic development

vs cultural heritage] became dominant as two national/local networks came into conflict. ... There was nothing inevitable about the action group's success and it was only because a particular set of associations were forged that the outcome took the form it did. ... In the course of collective action, resources will be mobilized in an attempt to cement the bonds which tie actors together (e.g. minerals for economic growth [vs] investments in parks and so on). (Murdoch and Marsden, 1995: 378)

Murdoch and Marsden's application of actor network theory provides both the analytical tools for conceptualizing the people involved and the relations between them. In the latter case, this theory supported the empirical findings that while obvious differences existed between different groups of actors working to oppose the development, nevertheless they were able to form a shared position and develop collective action (across a range of scales) on either side of the minerals extraction issue.

Looking at social actions more generally, a number of processes are often simultaneously involved in these situations. For simplicity's sake, they are separated out into four sets of activities, as shown in Table 9.2. Using the examples of constructing a gay identity in Hollywood (discussed in Chapter 7), and agitating for women's recognition and participation in Australian agriculture (discussed in Chapter 8), these four processes are summarized thus.

First, people experience circumstances that allow them to recognize difference and tension. As shown in the previous chapters, gay males in West Hollywood and women involved in farming have equally experienced forms of social and political marginalization. Forest (1995) argues that marginalization from citizenship was a stimulant for gay male interests in promoting West Hollywood's identity as a positive and creative gay district, while Liepins (1998a) notes that personal and industry-based exclusion from agricultural decision-making provoked diverse Australian women to form the 'women in agriculture' movement. Second, both geographers highlight the different social and power relations that contextualize these situations and provide activists with opportunities and arenas for action. Third, Forest and Liepins identify the strategic identities or subject positions which activists select and publicize (creative, responsible gay males and farmers/business-partners/mothers, respectively). In the Australian case, using a poststructural approach, I have concentrated on the opportunities provided by women's consideration of numerous subject positions:

The movement's exposure of multiple and contradictory subjectivities has provided women [with] political moments of choice ... subjects could realize agency and generate acknowledgement of a variety of knowledges based on the tensions and opportunities found in their particular positions and identities. (Liepins, 1998a: 152)

Finally (row 5, Table 9.2), through the cumulative awareness of experiences, contexts and identities, Forest and Liepins argue that social action is implemented through the choice of specific strategies and/or the construction

TABLE 9.2   GENERIC SOCIAL PROCESSES INVOLVED IN TWO EXAMPLES OF
SOCIAL ACTION

Social Processes	Constructing a gay identity (Forest, 1995) (See Chapter 7)	Seeking industry participation (Liepins, 1998a) (See Chapter 8)
People experience and recognize differences and tension	Marginalization of gay males as valued citizens.	Marginalization of women from agriculture as recognized contributors and able decision-makers.
People consider contexts of social and power relations	Gay males recognize: • the influence of the gay press • the legitimacy and economic symbolic opportunities of local authority incorporation.	Farm women recognize: • the legal and financial influence of farm partners • the economic and political influence of farmer and commodity organizations • the rural press • the legal, symbolic and political capacity of state and federal governments to recognize agricultural producers.
People consider different positions and construct/select strategic identities	A strategic (affluent) gay male identity is projected as creative, progressive, responsible and mature member of society.	A range of strategic subject positions are publicized and mobilized for the benefit of women and the broadening of agriculture, namely, farmer, business partner, mother, etc.
People consider and make choices	Press narratives of West Hollywood construct it as a place with a valid identity and structure as a gay community.	Individuals and groups within the 'women in agriculture' movement develop multiple actions, for example: • support of agricultural women • conferences and gatherings • media and newsletters • lobbying of industry groups and government.

and circulation of particular meanings. Forest argues that the construction of press narratives were crucial in projecting gay males as responsible and capable contributors to West Hollywood as a creative and design-oriented community (1995: 149–51). In a complementary fashion, I concluded that 'women in agriculture's strategies of personal support, public gatherings, media publicity, industry networking and political lobbying increased women's participation in farming at several levels:

In less than a decade, activists and supporters have seen growth in women's confidence and an insistence on being seen as farmers; they have encouraged greater participation of women in all aspects of farming (labor, administration, enterprise diversification, ownership and management); and they have forced an increased sensitivity to women's involvement in farming and industry decision-making. (Liepins, 1998a: 152)

While these types of generic process could be tabulated for many accounts of social action, it is important to note that these processes do *not* remove, or solve, social differences that occur within the groups under-taking the actions. Each case of social action discussed to date indicates various ways in which social and economic differences can potentially frac-ture the interest groups yet are temporarily set aside, or crossed in some fashion, so that collective goals can be achieved. For instance, differences between gay males or between farm women could have thwarted the goals and actions each group achieved in West Hollywood and Australia. A more graphic case of these differences illustrates how this can happen. As noted in Chapter 1, Valentine (1997) has reviewed the movement of lesbian separatism that developed in the USA in the 1970s (Box 1.2). She records lesbians actions in establishing and maintaining explicitly separate and alternative communities for lesbian lifestyles in rural USA. This is an example of the third field of action depicted in Figure 9.4, for these women were committed to a long-term alternative social environment involving the for-mation of land trusts and the development of women-only communities which aimed to foster non-hierarchical and self-sufficient lifestyles. In out-lining these activities, Valentine noted the women's purpose in 'separating from heteropatriarchal society', reflects the context of the 1970s radical feminism that had begun to identify 'heterosexuality as the root of women's (lesbian, bisexual and heterosexual) oppression' (Valentine, 1997: 110). While this was an ambitious movement, the character and success of the actions were affected by the differences between the women. The 'Lesbian lands' were sites that accentuated women's differences in ethnicity, class, sexual practice, and attitudes to boy children. She writes:

> [L]esbian separatist attempts to establish 'idyllic' ways of living in the countryside appear to have unravelled because, in common with tradi-tional white middle-class visions of 'rural community', attempts to create unity and common ways of living also produced boundaries and exclu-sions. ... Lesbian lands wanted to recognize and value diversity among the inhabitants of their communities, [but] the reality was that in many lands, identities were not equally valued, rather some were privileged over others. The [sexual, racial and class] tensions within these lands developed as a product of the inequalities of power and hierarchies that were constructed between the women. (Valentine, 1997: 118–19, 120)

In short, 'Lesbian lands' demonstrated the opportunity for social change, but also the many social differences and negotiations of identity and power that had to be navigated.

Reports of the tensions and fragility of action for social change are important. Social geographies could well be enriched if more continuous or longtitudinal studies were made of social action and the fortunes and negotiations of associated activist groups. This would occur especially where social geographers spent further energy on investigating how difference and collective beliefs, identities and actions were negotiated. Part of the challenge remains in the need to recognize, and find suitable conceptual apparatus to read the complex coincidence of social processes such as those identified in Table 9.2.

While tables might neatly divide components of social action, the material and cultural reality of most social action involves people in simultaneous and **overdetermined** social processes. Experiences, contexts, identities and strategic choices are all closely interwoven in different 'entanglements' (Sharp et al., 2000). These interconnected processes are not linear, but rather they variously mesh, grind or ignite in social conflict and/or adjustment in a variety of ways. Consequently, people are drawn into the need to interpret the contextual terrains of power relations surrounding them, and to choose specific identities and actions to fulfil their respective interests. Gibson-Graham's (1995) analysis of responses to changed work conditions in Australian mines is a case in point. As shown at the beginning of this book (Box 1.1) miners and their partners in Queensland and New South Wales had quite different responses to a new seven-day roster that mining companies introduced. Gender and class differences, together with different industrial histories and identities in the more conservative Central Queensland and more militant Hunter Valley locations, meant that contrasting actions were taken. Rather than a standard Marxist analysis, Gibson-Graham take a **poststructural** and **anti-essential** reading of these actions (see Part V for further explanation and contextual detail about this change in interpretation). The Queensland miners and their wives adopted the new work arrangements and used the higher wages to purchase local properties, take up other activities (farming and pig shooting) or buy investment properties on the coast while Hunter Valley workers and their partners resisted the changes with industrial action.

It is the spatiality of these actions that emerges when such contrasts are drawn. The spatial character of action results from the varying, context-specific ways space is encountered and shaped, and the way different places can become a medium through which differences are acknowledged but covered in the moments and activities of a specific form of social action. Thus the socio-spatial qualities and relations of social action require discussion.

9.3.2 Seeing social actions as socio-spatial phenomena

It may seem rather obvious to state that the distinction between geographic work and other academic accounts of social action rests in analysts' and authors' attention to the spatial character of such activities. Yet the socio-spatial qualities of individual and collective actions reported in this book are rich in the way they convey the social character of issues being contested. They are also qualities that are crucial to the effectiveness of the politics involved in each case, as will become apparent in this section.

In a general sense, social action has been shown in this chapter to involve the complicated negotiation of differences, identities and power relations. Rose (1999: 249) argues that 'space is a performance of power'. She notes that where social action involves the collective mobilization of different power relations and activities, then the performance of these actions will both take up and constitute/reconstitute space. In short, actions are not just about choosing to participate or not, or to maintain or change some aspects of one's individual or social frame. It also involves spatializations of the process. Delineating the issues, negotiating the surrounding terrains of power and implementing strategic actions *all occur in space.* Such space may involve symbolic sites, such as the Pentagon in Akatiff's (1974) example opening section 9.3, or it may involve the holistic construction of an idealized place such as the gay imagination of West Hollywood (Forest, 1995). Alternatively, this action-space may involve temporary sites, like those discussed by Winchester and Costello's (1995) analysis of the squats and hang-outs used by Australian street kids (see section 7.3.1). In all cases, however, the social processes constituting the actions (of protest, identity formation or the construction of marginal or alternative lifestyles) will be influenced by, and will shape, the spaces and sites in which the issues and people are positioned.

Murdoch and Marsden (1995) have termed this the 'spatialization of politics' and have noted the importance of understanding how relations occurring at different scales will be drawn together through different cases of local–national conflict. Parallel 'spatializations' can be seen in my reading of the 'women in agriculture' movement where activists mobilized across local, state, national and international arenas. They created relations with supporters and other organizations and positions that operated simultaneously across a number of scales (see section 8.3.3 or Liepins, 1998a; Panelli, 2002a). Both these cases illustrate what Massey (1999: 279) 'imagines' as space, being 'a product of interrelations ... constituted through a process [or continuous processes] of interaction'. Furthermore, in each of the cases introduced in this book, 'spaces of interrelation' could be identified and read.

Massey's imaginations about 'spaces of politics' are instructive for this review of social action. She presents two other imaginations of space, the second being where space is 'the sphere of the possibility of the existence of multiplicity' and the third being where space is 'disrupted and ... a source of disruption' (1999: 279–80). Both of these constructions of space can be seen in geographies of social action, and Fisher's (1997) study of alternative craft economies illustrates this well. Actions by these craft workers constitute the attempt to create long-term social and economic change in individual situations (part of the third field of action shown in Figure 9.4). Massey's space of 'multiplicities' can be seen in Fisher's description of how craft makers rework home spaces to achieve the work life to which they aspire. Home spaces involve multiple and often competing demands, people and functions. Kitchens, living rooms and other home spaces are simultaneously constructed as both productive and reproductive spheres – each with multiple facets.

It is also important to note that the social actions themselves further constitute space as a sphere of multiplicities. Fisher shows that craft producers' actions in creating an alternative economy for themselves could actively construct and delineate the multiplicities into their home spaces:

> Downstairs rooms often have spaces within them that are temporarily transformed in productive spaces. In these more indeterminate, temporary spaces, the boundaries between production and reproduction are necessary. What [craft] makers find desirable is an uninterruptible production space, yet at the same time an interruptible reproductive space. To achieve this, makers redraw boundaries between the two, by cordoning off such spaces with curtains and even bits of string, so creating permeable boundaries. Such membranes between the two filter out certain forms of interaction while allowing others. Pieces of string to mark out the space of production are reportedly among the most effective dividers between production and childcare: an invisible wall, allowing visual and verbal communication between maker and children, while in theory filtering out physical contact. (Fisher, 1997: 241)

Fisher's work illustrates the recognition of spaces as spheres of multiplicity, but she also demonstrates what Massey conceptualizes as 'disrupted space'. In the case of the Welsh craft producers, home spaces were clearly spaces of disruption and sources of disruption for workers who juggled multiple identities and responsibilities. Fisher writes of both the opportunities and frustrations of such space:

> [T]he space they work from can be empowering. In controlling the use of their space [craft], makers play a multiplicity of roles, creatively muddling conventional gendered roles and locations for the performing of these roles. However, makers often have to (re)demarcate spaces of production, 'roping off' work space from play space and creating often permeable, occluded spaces, the result being an ironic reference back to those delimited spaces of production and reproduction that they previously rejected. (Fisher, 1997: 247).

The complexity of attaching multiple activities and relations to various spaces also involves both territorial and imagined arenas. Gibson-Graham's recent work analysing capitalist and non-capitalist accounts of community and regional economies is a case in point (see Part V)

(Community Economies Collective, 2001; Gibson-Graham, 2003). Working with regions defined in capitalist discourses as 'declining' or 'marginal', Gibson-Graham have partnered local groups to record alternative non-capitalist economic and social options that can operate in people's every-day lives. These projects in Australia and the USA have enabled the imagination of different, more positive, local and regional identities as well as specific material household arrangements.

Further contrasting examples demonstrate the physical and imagined spatialities of social action. Routledge's (1997b) account of protest of motorway construction, and Gibson's (1998) analysis of Aboriginal music and indigenous self-determination document space as interrelational, as multiple and as disrupted. Routledge's account of resistance to the con-struction of the M77 motorway illustrates the second field of action in Figure 9.4, whereby protesters mobilized an alternative, non-violent 'events action' strategy to communicate their opposition to the develop-ment of the motorway. The range of participants show how action occurs across many differences as local residents, councillors, students, profes-sionals, children, unemployed and long-term environmental and peace activists joined to engage in discursive and physical struggles over the motorway issue. The action is one Routledge identifies as a symbolic 'post-modern politics of resistance' for although protestors were engaging with Scottish motorway and regional council issues, they were also mobilizing activity and performing their protest for regional and national arenas. Spaces thus became integral to the struggle. Not only was the physical space of the planned motorway involved, but also the wider media arenas were engaged in order to construct an imagined 'Pollock Free State' that could disrupt the mainstream politics and economics of motorway develop-ment. Here, then, a physical location becomes a space of multiplicity and disruption in both creative and radical ways.

Multiplicity occurred since the motorway would cut through an estate south of Glasgow which 'Sir John Maxwell of Pollock, founder of the National Trust of Scotland, bequeathed ... to the citizens of Glasgow' (Routledge, 1997b: 364). This space had contrasting values and therefore the tactics of the protestors drew on this conflict to create the 'Pollock Free State' as an 'ecological encampment' that could convey a radically dif-ferent environmental discourse and construction of space from that of pro-posed motorways (see Figure 9.7). Routledge explains:

[T]he camp acted as a visible symbol of resistance to the motorway. There was a continually fluid component to the camp in terms of its architecture and its residents. The Free State was in constant flux, changing face and form continuously as tree houses were completed, benders deconstructed and rebuilt, and as new inhabitants arrived to build their homes and others left for various periods of time. ... The Free State represented the 'home-place' and the focus of the resistance against the M77, articulating an alter-native space that occupied symbolic and literal locations. (1997b: 365–6)

FIGURE 9.7   POLLOCK FREE STATE TREE HOUSE
Source: P. Routledge, private collection.

As a space of disruption, the Pollock Free State could draw on alternative discourses and provide a physical and symbolic site for the diverse participants to meet and join across their differences to enact protests against the road building.

In a different and even more long-term struggle, Colin Gibson (1998) presents a geography of social, cultural and political action for indigenous self-determination in Australia. This illustrates a further example of the third field of action depicted in Figure 9.4, whereby action is established by people wishing to achieve a long-term change in Australian society. Taking the example of contemporary Aboriginal music, Gibson shows that indigenous groups and individuals are creating an alternative politics of music spaces that are creating a social, economic and political impact. In Melucci's terms, this action forms a social movement of solidarity that seeks to breach the current Australian socio-political context and – through music – create different values and political rights in terms of self-determination.

Gibson argued that indigenous artists (writers, producers, sound engineers, roadies and performers) come together across occupational, gender and ethnic differences to form networks of action that can change the meanings and spaces of popular music production and political thought. He explains:

> The medium of popular music is central to the construction of arenas in which individuals with few resources and limited access to mainstream economic and educational opportunities can participate – developing skills and fostering community pride for their expression. Furthermore, popular music offers Aboriginal people a language and medium, which is often more easily received by mainstream Australian society than formal political discourse. ... In an era where conventional regimes of nation and identity are being destabilized, popular culture, in this case popular music, becomes a crucial public sphere of geopolitical change. (C. Gibson, 1998: 165)

Gibson shows that a soundscape is being constructed that disrupts the dominant Australian discourse about land, nationhood and the position of indigenous people. This alternative musical activism promotes a land-based identity and case of self-determination. Through a variety of music spaces this alternative politics for social change is enacted. The spaces include micro-sites of song lyrics – power texts narrating land-rights, self-determination, survival and resistance (see Table 9.3).

TABLE 9.3   RECLAIMING LAND AND SELF-DETERMINATION:
SONG LYRICS AS SOCIAL ACTION

Rights to land and self-determination	Cultural survival
'Fight for Your Land' (Titjikala Desert Oaks Band, 1989)	'We Have Survived' (No Fixed Address, 1981)
'The Land and My People' (Wirrinyga Band, 1990)	'Cannot Buy My Soul' (Kev Carmody, 1991)
'Land Rights' (Sunrize Band, 1995)	'Forgotten Tribe' (Coloured Stone, 1994)
'Mabo' (Yothu Yindi, 1995)	'Proud Young Arnhem Land Man' (Wirrinyga Band, 1996)

Source: excerpts from C. Gibson (1998: Table 1, p. 175).

The soundscape of indigenous popular music includes the alternative recording, performance and distribution spheres that counter hegemonic, mainstream, popular music production and distribution arenas that exist in the Australian urban – and internationally linked – music industry. In particular, Gibson highlights the community association with many bands and the performance sites that provide alternative venues and access points. Consequently, he concludes:

[T]he emergence of Aboriginal popular music in Australia has thrown up significant challenges to the nation's geopolitical legitimacy, histories, assumed national identities and social landscapes. ... The creative process involved in writing, recording, and performing songs provides an aspect of individual empowerment; and new networks of Aboriginal-controlled production, broadcasting, distribution, and marketing provide some resistance to the hegemony of music multinationals and a legitimacy to performance activities. These networks and the matrix of tours, festivals, and performance events (which expand every year), inscribe distinctly Aboriginal social spaces onto the Australian musical landscape. ... They provide their Aboriginal and non-Aboriginal audiences with a powerful expression for the survival of indigenous peoples and distinct postcolonial geopolitics. (C. Gibson, 1998: 180)

Considering the multiple scales at which people may act ensures that social geographies continue to be written, showing that social action is possible throughout the material and imagined spaces we can recognize. While other disciplines may focus on the social sources and relations mediating action, geography can provide accounts of the mutually constituting social/spatial entanglements that provide a possibility for meeting across difference to chose actions, and in some cases broad social and political change.

9.4 CONCLUSIONS

This chapter has formed the final example of how social geographers can simultaneously acknowledge social differences yet construct valuable knowledges about how people work and act across those differences. In this case, social action has been the focus, for hopeful yet reflective geographies have been written to show how people will join across various differences to build lifestyles, places and protests that involve some degree of shared meanings and relations. In this chapter we have seen how great a range of social actions can be identified. It has also been important to acknowledge that all social actions 'across difference' involve a range of social processes by which people make meanings of their experiences (or grievances); identify relevant power relations and positions (or identities) from which to act; and implement specific actions to achieve their ends. Furthermore, after Massey (1999), the spatial qualities of social action have been illustrated as key arenas where actions are performed (in relational spaces) even while differences can be both sustained (in spaces of multiplicity) or capitalized upon (in spaces of disruption).

Geographies of social action are often invigorating and provocative to read once these generic social and spatial qualities are noted.

Renditions of social action are also important for the ongoing practice of social geography for at least four reasons. First, geographies of social action remind us of the rarely fixed, but rather dynamic and spatial dimensions of social life. These studies emphasize that it is insufficient to create social geographies of patterns and description if we do not also acknowledge the constantly changing and negotiated nature of social life.

Second, geographies of social action highlight the spectrum of differences that shape our social worlds and the knowledges we produce. To act is usually to begin with some issue about difference: self and other; your experiences/lifestyle/opportunities/needs compared to others. Thus geographies of social action can integrate the complex relations between multiple differences and the possibility of connections across those differences. While connections across difference may be studied through attention to households, neighbourhoods, communities and nations, attention to difference and/within social action challenges geographers to construct understandings of 'why' (and how) some form of unity can be achieved across difference in often transitory ways.

Third, geographies of social action raise the problems and challenges constantly faced by social (human) geography and social sciences in choices we make about how to write and re-present social life (in this case social action). Geographies of social action encourage us to consider our own multiple interests and agendas. The perspectives, theories and constructions geographers use to communicate analyses of action create particular and situated knowledges based on both the geographers' research choices and their own experiences and beliefs about what social geography can be and how it should be practised. This situation is not limited to studies of social action, yet the usually volatile and explicitly political nature of much social action magnifies the need for social geographers to present clear and reflective accounts of both the actions and their constructions of it.

Finally, geographies of social action remind us of the potential and depth in social geography, for in constructing geographies of socio-spatial life we constantly face the possibilities (hopes and anxieties) surrounding our choices, capacities and abilities to shape and change our social worlds. These possibilities may be conventionally explored through existing structures and relations or they may stretch along a continuum of critical thought and advocacy for social change. This then brings us full circle in this text – back to the questions closing Chapter 1 and the theoretical perspectives of Chapter 2. We have moved through the studies of diverse people and the relations and spaces with which they engage to re-engage with the motivations, theoretical approaches and re-presentations we can select in writing social geographies as positioned and powerful/power-filled geographies of difference and action.

SUMMARY

- While attention to social differences, identity and power relations enable geographers to consider how groups of people and societies are constituted in different places and space, it is social action that highlights how dynamic societies and individual experiences can be.
- Geographers study a range of types of social action from individual acts through informal and semi-structured group activities to formal political strategies.
- Social action can also be distinguished by the time frame and social purpose of the action. The purpose or motivation of the action can be understood in terms of the degree to which action is seeking to maintain or breach a society's codes and structures (Melucci, 1996).
- Theoretically, a wide range of perspectives can be used to conceptualize the contexts, purposes, actions and people involved. Commonly adopted approaches in contemporary social geography include *actor network theory* and poststructural concepts of *political subjectivity*, the former focusing on the networks and processes by which an issue is identified and responses made, the latter highlighting the way people can recognize – and use/reconstruct – the way different discourses position them and define the relations and spaces with which they want to engage.
- Some geographers trace the contexts and relations (and networks and spaces) that underlie the social processes involved in action as people highlight or seek to mobilize or stretch across differences to achieve some preferred outcome.
- Geographers also focus on the ways social and spatial dimensions of action are entwined – whether this be specific physical spaces of a home that is delineated for different functions or of whole regions and soundscapes that are imagined in alternative configurations to provide options in economic and cultural expression.
- While action can often be studied as a positive, transformative and unifying phenomena, it is important to remember that strategic identification and action under common goals or identities will not eradicate differences but can reconfigure or temporarily set them aside in various performances and connections.
- Finally, geographies of action are effective in reminding us that social geography is also crucially predicated upon the questions we choose to pose, research agendas we set, practices we adopt and representations we elect to make of our social world.

Suggested reading

Reading about social action can take at least two forms. Theoretical pieces provide a conceptual foundation for how action is observed and analysed while case studies of social action give everyday examples of specific activities and settings. On *structuration* theory, it is useful to look at Giddens' (1984) comprehensive book and geographers' use (e.g. Moos and Dear, 1996) and critique of his approach to action within a structure/agency framework (e.g. Chouinard, 1997; Gregory, 1989; Gregson, 1989). On *actor network theory*, starting points are provided by Law's book (1994), while geographers' applications and review of this theory are worth browsing (e.g. Murdoch's (1997) 'think piece' and the special issue of *Environment and Place D: Society and Space* (see Hetherington and Law, 2000)). On *social movements* and *identity politics*, books by McAdam et al. (1996), Melucci (1996), Crook et al. (1992), Keith and Pile (1993), and Pile and Keith (1997) give contrasting perspectives on the character and operation of strategic collective action in a variety of settings, and access to geographical accounts of how people strategically operate on both individual and group issues are found in the case study articles cited in this chapter (especially Dwyer, 1999; C. Gibson, 1998; Routledge, 1997b). Finally, in terms of how geographers approach their own work as social action, there is a range of creative and reflective pieces that can inspire. For example:

- Bunge (1977) presents an alternative account of social and 'expeditionary' geography during the emergence of radical geography.
- Various journal articles contain reflections on how academic geography may include activism – Blomley (1994), Castree (2000) and Tickell (1995).
- Gibson-Graham (1996b, 2000) provide accounts of how their work may produce alternative geographies and options for the people with whom they conduct research (Community Economies Collective, 2001).

CONCLUSIONS AND OUTLOOK

PART IV

All closures are difficult, frequently because they are either abrupt or must be contrived for various practical or rhetorical reasons. This ending is no different. In drawing this book to a close I reflect on the ground we have covered through the chapters which have provided an overview of contemporary social geographies. But it is also important to draw links between and across the structural divisions that the chapters present. The result is a simple synthesis that may provide a framework for continuing encounters with social geography and the topics it can usefully investigate. Finally, this conclusion operates not so much as a closure but as a gateway to future geographies and an invitation to continue documenting, questioning and challenging the constitution of our social lives and worlds. In the final parts of Chapter 10, I highlight ways to recognize and read some of the recent debates, and I suggest some reading strategies for keeping up to date with new directions in social geography.

10 Conclusions and Future Directions

10.1 A REVIEW

This book has been written out of a course in social geography where students found some of the best parts of the subject involved the connections and links that could be made across the material at the end of the course (see Box 10.1). This conclusion seeks to achieve a similar result.

Box 10.1 Reflecting back over social geography: best parts of the subject?

... broad canvassing – it helps you look at society in a different way.

... the way it made me really think about the issues. It changed my opinion on a number of things – has made me view things such as space (as heterosexual) differently.

... the way in which everything throughout the subject comes together at the end.

... near the end I started making connections between different parts of the course.

... the end when we could make connections between the issues, subjects and theories.

... being provided with the opportunity to tie all the key ideas together.

Source: Students' Course Feedback, 2002.

This book has sought to outline how geographers acknowledge social differences and the uneven ways these differences are played out over space – through everyday life, through specific choices, and sometimes through broader socio-political struggles. But recognizing difference has not been sufficient for most scholars. Instead, they have concentrated on the geographic contexts and consequences of this diversity, noting how uneven relations and performances of power are involved, and how various identities and actions are assembled and implemented, sometimes because of, and other times across, those differences.

Chapter 1 began by illustrating some of the experiences, conditions and actions that have been the subject of social geography in recent times. It highlighted the fact that these geographies have concentrated on the identification of social difference and action, and acknowledged the unevenness and struggle that has sometimes followed in specific places. In order to map, explain, critique or make readings of these differences the chapter closed by noting that social geographers have undertaken their research and writing by working from 'somewhere' (section 1.1). This **positionality** (whether explicit or implied) means that, to understand contemporary social geographies, we must have a grasp of the scientific, philosophical and theoretical approaches that have underpinned much of this work. These matters formed the substance of Chapter 2. In this discussion of 'negotiating science, theories and positions', the examples of Paul Cloke and Geraldine Pratt illustrated how geographers find themselves responding to different academic environments, theoretical developments and research topics. A sense of social geography was given as a dynamic activity where layers of experience and theory are built up over time. More broadly, much social geography has moved from a quest for spatial scientific patterns, laws and solutions to modern problems (e.g. ghettos or prostitution) to one that concentrates on multiple interpretations, sometimes radical critiques and, most recently, diverse readings of the way society and space mutually constitute each other through material and discursive processes.

The second part of this book then turned to a series of detailed commentaries of the four broad axes of social difference that have structured social geography over the past four decades. While each chapter on class, gender, ethnicity and sexuality had a distinctly different focus, they all demonstrated the way the same broad theoretical approaches had been applied to research. They demonstrated that each category of social difference not only produced spatial manifestations but that these were mutually constitutive and conveyed the existence of uneven power relations, whereby one form of category (e.g. whiteness or heterosexuality) might be more privileged or powerful than another. They also showed how recent studies of single categories of social difference necessitated a greater awareness of the connections and interlinkages with other axes of difference, making a more complex but also more nuanced account of how social differences intersect for various individuals and groups, and in contrasting places and spaces. For instance, in closing Chapter 3 it was noted that understandings of class must now recognize that any given class category will be 'in process' and **overdetermined** or mutually connected with other social dimensions. Thus a single class can be recognized as differentiated by ethnicity, gender and over space (McDowell, 2000). In a similar way, Chapter 5 showed how experiences and negotiations of ethnicity are also transected and shaped by other differences such as class, gender, religion and age, which are played out in different ways in contrasting environments and specific places (e.g. Dwyer, 1999; Rees et al., 1995).

The coexistence of multiple social differences and the dynamics of social life (i.e. how people do not simply live static lives within social categories) has resulted in geographers' attention to matters beyond the definition and critique of these categories of difference. Consequently, the third part of this book concentrated on how geographers have studied social relations, choices and actions that are predicated on strategic use of these differences, or strategies that stretch across such diversity. Chapter 7 showed how notions of identity can be used to understand people's selection and articulation of some aspects of social difference over others, or alternatively, how they have to negotiate the consequences of other groups or institutions structuring aspects of their lives as a result of association with a certain category (e.g. woman, homosexual, etc.). This work illustrated how identity can be strategically formed for individuals, groups, locations or whole nations. The processes of identity formation or contestation of dominant identities raises further issues, specifically to do with concepts of power and action.

In Chapter 8, geographies of power were discussed, including the way social meanings and relations are conveyed through various structures and processes of power. Examples of state, industrial and media arenas illustrated how the power relations underpinning social differences are negotiated in different settings where dominant meanings and arrangements were shown to be powerfully positioned in certain buildings, procedures or discourses. But entangled within these circumstances there are also points, opportunities and sites through which domination or hegemony can be challenged, modified or even directly countered (e.g. the informal *mteto* relations operating within the class- and ethnicity-based strictures of the South African mining compound – see section 8.3.2). Space can be understood as 'a performance of power' (Rose, 1999: 249). At times, specific sites (from bodies, through homes, to workplaces and wider social and imagined spaces) become the subject of struggle and multiple meanings and performances of power. Indeed, at micro scales, geographers have also come to appreciate how individuals occupy multiple positions as subjects within different power relations and spaces. As Radcliffe notes:

> [T]he very positionalities of subjects – individuals' own relationship to power relations – rests in part on their insertion into the social landscape of power. In other words, the social markers of difference – whether gender, age, race or nation – around which people identify and live their lives are in themselves markers invested with power, reinforced through the representational and discursive constructions of difference and opposition … [and] the mapping out of difference onto place and landscape through the spatially defined daily routines of social groups and individuals … means that social difference acquires its own spatiality and representational power. (Radcliffe, 1999: 238)

Case studies in Chapter 8 showed how different groups of individuals are positioned in state, industry and wider landscapes, but these works also heralded the importance of appreciating people's choices and action. Thus,

211

Chapter 9 outlined how social difference can be converted into action. Whether as a response or challenge to social differences, or as a strategic use of key differences, geographies of social action show how people can use identities and engage in power relations to realize particular goals or outcomes. Geographies of social action also show the significance of space in a myriad of forms, including spatial contexts, processes and outcomes. Indeed, action always originates from somewhere and employs space in a complex of material and symbolic ways. Individual actions around work and home arrangements illustrated by Fisher (1997) remind us of how material and imagined home/work spaces coincide and act as possibility and constraint for home-based craft workers. More broadly, Murdoch and Marsden's (1995) mining protest study and C. Gibson's (1998) reading of indigenous Australian music showed how the spatialization of politics can operate through local to national networks and scales. The possibility presents itself, then, for difference, identity, power and action to be drawn together.

10.2 SYNTHESIS AND FUTURE DIRECTIONS

While this book has highlighted the individual core elements of social difference, identity, power and action as they relate to social geography, attention to the relations between them enables a basic synthesis to be made. Figure 10.1 portrays one such synthesis. It presents a reading of social geography where social life in all its diversity is acknowledged as situated in – and reproducing – spaces and places (central to the figure).

A social understanding of space, in non-essentialist terms, will not precede society but will be 'filled, contested, and reconfigured through contingent and partially determined social relations, practices and meanings' (Natter and Jones, 1997: 149). Such views of space and place not only avoid fixing dimensions or meanings of space, but also remind us of the constant making and remaking of space and place that occurs as we live our day-to-day lives (Massey, 1999). Indeed, Massey has argued that interrelational and open notions of space better support our endeavours in studying both the dynamic, changes and the political possibilities of space:

> [I]magining space in terms of interrelations and of constitutive difference can provide one of the preconditions for this openness, for the possibility of the emergence of genuine novelty. Both the endless openness of spatiality (its loose ends) and its inherent disruptedness (its conflicting co-existences, its unexpected distancing, its dissonant or congenial juxtapositions ...) establish the grounds for such newness. Imagining space in terms of multiplicity and the possibility of interrelations guarantees the openness of the outcome of any interaction (or lack of interaction). The space is neither stasis nor closure. ... This 'spatial' is the very product of multiplicity and thus a source of dislocation, of radical openness, and so of the possibility of creative politics. (Massey, 1999: 287)

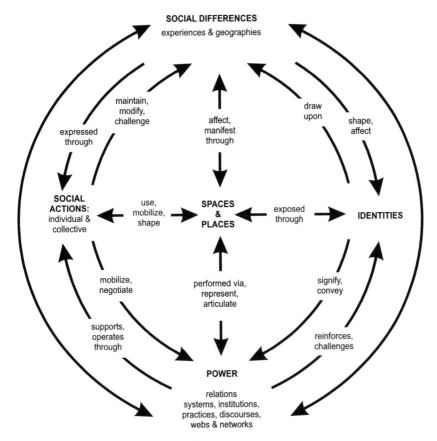

SOCIAL DIFFERENCES
experiences & geographies

maintain,
modify,
challenge

affect,
manifest
through

draw
upon

shape,
affect

expressed
through

SOCIAL
ACTIONS:
individual &
collective

use,
mobilize,
shape

SPACES
&
PLACES

exposed
through

IDENTITIES

mobilize,
negotiate

performed via,
represent,
articulate

signify,
convey

supports,
operates
through

reinforces,
challenges

POWER

relations
systems, institutions,
practices, discourses,
webs & networks

FIGURE 10.1   A SYNTHESIS OF SOCIAL GEOGRAPHY

These perspectives on space and place encourage attention to inter-connections with the other core elements under discussion in this book, and these linkages operate with all four of the other elements in Figure 10.1.

First, moving to the top of the figure, we can appreciate that social differences are experienced unevenly through space. These differences constantly reproduce places, reinforcing certain categories of difference as normal while others are marginalized or controlled (e.g. women partici-pating in mainstream farming organizations, or gay and lesbian affection in public streets or restaurants).

Second, directly across from social difference, at the bottom of Figure 10.1, power is portrayed as equally linked to each element. These links occur as power relations are implicated in the uneven experiences of social differ-ences; power is represented and articulated through spaces and places; and, as will be shown below, power is directly involved in the performance of identities and the operation of social actions.

Then, to the right-hand side of Figure 10.1, we note that social differences shape and affect the formation or practice of identities that are associated with individuals, groups or places. Conversely, when identities are being strategically assembled or even challenged and reconstituted, they draw up specific aspects and understandings of social difference (e.g. as shown in section 7.3.2, national identity formations in Finland or New Zealand drew on selective differences associated with gender, class and ethnicity). These identities convey certain power relations that privilege dominant categories of social difference (e.g. masculinity and whiteness) and they are expressed through varied spaces, ranging from the imagined sphere of a nation to the specific sites of symbolic places, landscapes, even buildings and bodies.

Finally, to the left-hand side of Figure 10.1, social action is depicted, again in a mutual relationship with social difference and power. Social differences may be highlighted through actions that in turn may be established to maintain or even challenge those differences (e.g. the rearticulation of a specific ethnicity through themes of indigenous self-determination in Aboriginal Australian music as shown in section 9.3.2). In addition, practices and webs of power will underpin and operate through social actions. This may occur by negotiating with specific institutions or relations of power (e.g. the dominant commercial music networks in Australia). In producing these social actions, spaces will be invoked and mobilized. In addition, specific places of significance may be promoted or even reconstituted as they form a terrain from which people can act and shape their lives and social worlds.

This synthesis provides one snapshot of contemporary social geography. The connections and interactions depicted here have been explicitly arranged to highlight the dynamic spatiality of social life – in all its material, cultural and political forms. As noted in section 1.2, this book has been written from a composite of explicit interests and this synthesis is no less a product of those interests. The critical, poststructural and non-essentialist approaches that I have used in research to date lie behind this snapshot. The figure thus attempts to portray some of the mutually connecting and overdetermined qualities of social life and the geographies that might be conceptualized.

But multiple linkages and mutual constitutions are not the only motivations for the content of this book, nor for many of the contemporary social geographies being written today. Instead there is also a continuing critical and political component that inspires the subdiscipline. Social geographies can contribute to what Massey argues as the 'spaces of politics': 'a reformulated politics and space revolves around the openness of the future, the interrelatedness of identities, and the nature of our relations with different others' (1999: 292). While this reformulation is relevant more broadly throughout human geography, social geography provides fertile ground for such practice. Returning to the questions at the conclusion of Chapter 1, the material in this book has highlighted how a range of approaches can support the continuing identification of inequalities. The biographies included in Chapter 2 and Part V remind us that in each different era of geography particular theories and practices have dominated the 'way things are done'. The biographies have been particularly explicit in showing how geographers' contexts have influenced the forms of knowing that have prevailed in various academic environments. These frames of theoretical and research practice have shaped what is valued or critiqued in the social world and the research environment. However, Figure 10.1 and Massey's revisioning of 'spaces of politics' suggests that beyond the identification of difference, key questions regarding inequalities, power and the practice of identities and strategic actions will continue to be important in social geography inquiries.

While postmodern, poststructural and cultural interests have provided core energies in the writing of social geography over the last decade, there is also a continuing and reinvigorated commitment to the action, politics and material dimensions of our work. This has taken a number of forms that together provide five directions for future work.

FURTHER DIFFERENCES

The importance of recognizing social difference continues to infuse social geography. At times this is a case of being attentive to the heterogeneity and inequalities that characterize a particular social group that has been the focus of study for some time already. For instance, Geraldine Pratt's engagement with Filipino perspectives have further enriched her work on gender and employment (see Box 1.3). At other times, continued attention to difference is producing new bodies of work on other axes of difference beyond the traditional foci of class, gender, ethnicity and sexuality. These fields include geographies of disability, of youth and age, of homelessness, and of travellers, to name but a few. In some cases, a substantial literature now exists, but in each example we re-meet the challenges of reading social-spatial connections and the uneven power relations and possibilities for action that may present themselves in the lives of specific individuals and groups (see some examples in the work of Cloke et al., 2000; Davis, 1997; Gleeson, 1999; Holloway and Valentine, 2000; MacLaughlin, 1998; Mattingly, 2001; Panelli, 2002b).

THE MATERIAL AND THE EVERYDAY SPACES OF LIFE

The contexts, and even mundane settings, in which we conduct our lives at home and work, have continued to be of interest to social geographers. These places are recognized as important, for the negotiation of social differences such as class, sexuality or age are not simply conceptual social debates or cultural struggles; they are embedded in specific sites and material experiences (Jackson, 2000). Social geographies will continue to be written about how people live with (and sometimes across) difference in ways they use – and sometimes reshape – the everyday spaces of their lives. These will include the layered material and socio-cultural spaces of sites known as 'home' and 'work', but they will also increasingly incorporate deepening analyses of other spaces: margins, thirdspaces and cyberspaces where new or reworked understandings of space can be further developed (Gough, 2001; Kitchin, 1998; Mowl et al., 2000; Panelli et al., 2002; Soja, 1999; Valentine and Holloway, 2002).

THE PERFORMED

Drawing on a range of recent theory, social geographies will increasingly document the practice of social life. In some cases this may be conceptualized in specialist theoretical terms through such notions of **performativity** or **non-representational** theory. However, whatever the perspective employed, some geographers will modify the popularity of cultural theories and highlight that our subject and the societies and individual lives we study are 'enacted' and can be thought of as a 'flow of events' – not just a cultural construction or representation (Barnes, 2002; Harrison, 2000; Longhurst, 2000a; Rose and Thrift, 2000).

THE POLITICAL AND ETHICAL

The political dimension of radical geographies continues to be relevant for social geography and is increasingly combined with reflections on the practice, methods and **ethics** of our research. While a fascination with postmodern and cultural approaches to geography can be questioned, social geographies can continue to highlight difficulties surrounding social relevance and inequality (Hamnett, 2003; Martin, 2001). For example, social geographies can combine opportunities for action, activism or policy development on the part of the scholars involved (Blomley, 1994; Pain, forthcoming). In a different form, Paul Cloke's (2002) recent writing includes a questioning of the ethical and political values and choices geographers can make in their work. Further questions on morality, ethics and the purpose of social geographies is likely to increase (Aitken, 2001; Gleeson and Kearns, 2001; Martin, 2001; Proctor and Smith, 1999; Roberts, 2000; Smith, D.M., 1999).

NEWER TERRITORIES – EMOTIONS AND SPIRITUALITY

Finally, as a result of the ongoing interest in the performance, politics and ethics of writing social geographies about difference and inequality, some authors are beginning to look at new dimensions of society and the geographies that have been absent to date in much of the subdiscipline. For instance, building on from earlier humanist and feminist geographies, some authors are arguing that emotions (a key set of social relations) should be more explicitly included in social geographies as both a subject and a theme for researchers' reflexive thought (Anderson and Smith, 2001; Widdowfield, 2000). In a similar way, other geographers are considering the role of values, religion and spirituality in either their research and/or the practice of their geography (Cloke, 2002; Holloway and Valins, 2002; MacDonald, 2002). These new directions may provide further tools and perspectives for a continuing openness to the critique of societies and the power relations and actions that shape them in different settings.

SUMMARY

- Social geographies recognize social differences that are unevenly experienced across society and space.
- Geographers use a variety of contrasting approaches to study these socio-spatial phenomena.
- Different places (from individual bodies and homes, through to workspaces, urban neighbourhoods and rural communities) form settings in which multiple social differences occur at any given time within a society.
- Social differences are interconnected with each other (e.g. sexuality transected by gender, ethnicity and class and so forth) but in different spatial manifestations.
- The meanings and uneven experiences associated with social differences result in people negotiating difference through identity choices, power relations and social actions.

Suggested reading

As social geography continues to develop as a topical and multifaceted sub-discipline of human geography, progress and trends can be traced through a variety of further readings. One step can involve following up many of the authors listed in the last section of this chapter, where the five current trajectories were identified.

A second step can be taken through a continuing reading of the core social differences that form cleavages in both society and space. This is often most usefully achieved by browsing recent issues of selected journals. Scholars writing on difference and action publish in a wide range of arenas so it is always useful to consult the contents pages and internet homepages of: *Annals of the Association of American Geographers*, *Area*, *Journal of Social and Cultural Geography*, *Progress in Human Geography*, and *Transactions of the Institute of British Geographers*. Also, for class, see *Antipode*, and *Environment and Planning A*; for gender, see *Gender, Place and Culture*; and for ethnicity and sexuality, see *Environment and Planning D: Society and Space*.

Finally, to consider the range of recent research and debate that informs social geography it is always informative to read the articles and progress reports that are regularly published in *Progress in Human Geography*, including those explicitly on social geography (e.g. Pain, forthcoming), as well as associated relevant topics such as ethics, ethnicity and race, gender and geography, qualitative methods, quantitative methods, rural geography, sexuality, urban geography and so forth.

INDIVIDUAL SOCIAL GEOGRAPHIES

PART V

Andrew Herod: Questions of class and geographies of labour
Accompanying Chapter 3

Jo Little: From 'adding women' to reading the 'practice and performance' of gender
Accompanying Chapter 4

Kay Anderson: Disturbing 'race' – its genealogies and power
Accompanying Chapter 5

Larry Knopp: From 'mainstream radicalism' to explicit challenges and contradictions in (sexualized) geography products and practice
Accompanying Chapter 6

Karen M. Morin: Developing poststructural geographies of subjectivity and identity
Accompanying Chapter 7

Michael Woods: Developing an eclectic critical engagement with theories of power
Accompanying Chapter 8

J.K. Gibson-Graham: Creating geographies of action – collaborative and alternative
Accompanying Chapter 9

Andrew Herod is Professor of Geography at the University of Georgia and his work demonstrates the ongoing relevance and political capacity of research that focuses upon class. While many geographers have engaged with other important social categories, such as gender, sexuality and identity politics in recent years, Andrew's work shows that class continues to be a crucial difference that people daily negotiate on many scales, one that geographers can continue to work with effectively and critique in a range of political and academic projects.

Perhaps the awareness of class as a key form of social difference is strong because Andrew remembers coming to understand some class issues well before entering university. He recalls:

> As a child I had grown up listening to my grandfather, who had been very involved in the labor movement in Britain, talk about how union negotiations and workers' activities had shaped patterns of investment in the British steel industry, so deep down I knew that workers and their organizations were important actors in shaping the landscapes of capitalism. (Interview, 2003)

A more formal sense of class, and the unequal capitalist society in which it operates in Britain, came later as Andrew remembers learning about 'welfare geography' in secondary school and hearing about the ideas being published in the late 1970s and early 1980s in books such as D.M. Smith's *Where the Grass is Greener* (1979) and David Harvey's *Social Justice and the City* (1973).

Andrew's own interest in questions of class was kindled as he took up undergraduate studies at the University of Bristol.

> I took several classes on economic and urban geography, together with a class on the philosophy of science, which got me thinking about how landscapes were shaped by broad economic and political forces. In my final year at Bristol I took a class called 'Marxism in Human Geography' taught by Keith Bassett. This allowed me to put together theoretically ideas about

economic and political processes and ideas about landscape formation and to understand how landscapes are shaped and reshaped by processes of capital accumulation.

Much of this account may be similar to many other students' experiences. However, Andrew converted this interest into action as he investigated the impact of industrial restructuring on workers. Studying at West Virginia University, he investigated the class relations and politics associated with the closure of a glass manufacturing plant (see Herod, 1991a and 1994). This initial research came from his approach to class, which was 'steeped in the political economy/Marxist tradition, to which I had been initially exposed as an undergraduate at Bristol University in England in the early to mid 1980s' – much like that described in section 3.3.2. Things were to change, however, for Andrew became interested in how geography (especially scale) affected class processes. Influenced by Neil Smith's consideration of scale, Herod (1991b) came to understand class and scale as closely linked in geographies of labour. In contrast to much of the Marxist geography of the 1980s that had 'focused almost exclusively upon the actions of capital', Andrew developed an analysis of how class could be studied through an investigation of labour relations and their impact on space and scale (Herod, 1991b), the idea being that capital was not the only player in capitalist societies, or the only force in creating capitalist landscapes. Rather he viewed 'workers and their organizations as [also] important actors in shaping the landscapes of capitalism [and] began to think about how workers' actions could be better incorporated into theorizations of how the landscapes of capitalism are made' (Interview, 2003).

This goal produced the beginnings of the 'labor geography' for which Andrew is well known. He commenced with a study of the US East Coast dockers' union (the International Longshoremen's Association) for his PhD (at Rutgers University) that resulted in a range of material arguing that workers – as a series of diverse labouring classes – can have an integral part in shaping the relations and spaces/landscapes of capitalism (Herod, 1992, 2001a). He explains:

[In] my work on labor I have sought to show that different groups of workers may have quite different spatial visions for how they would like the geography of capitalism to look. Consequently, there may be quite significant cleavages within groups of workers. ... A sensitivity to such varied spatial visions helps 'deconstruct' the black box of class to show that the category of 'labor' (or of 'capital') is not monolithic, that not all workers will adopt similar political strategies even if, by definition, they share some common class characteristics (e.g., none of them owns the means of production). Rather, different workers will adopt different positions on particular issues as a result, perhaps, of their geographic location – are they living in a part of the country which is booming economically or which is in recession? – or of their local traditions of militancy and/or political acquiescence. By adopting a specifically spatial approach towards issues of class, then, I hope to allow for a more sophisticated and nuanced reading of 'labor' and 'capital' as class categories. (Interview, 2003)

Andrew's recognition of diversity within class categories has corresponded with the growth of poststructuralist thought and the approaches taken by geographers such as Gibson-Graham (see section 3.3.3). He notes that attention to the role of discourse in the production of class categories and capitalist rhetoric have been important recent advances in his thinking.

I have become interested in the writings of some of the poststructuralists and so have explored the ways in which categories like class have been constructed discursively. ... [In] one paper I wrote concerning inter-union struggles over access to work on East Coast waterfronts (specifically, which union – the ILA or the Teamsters – would get to do this work), I was particularly interested in the discourses used by the rival unions to bolster their claims to the work. (Interview, 2003; see also Herod, 1998a)

Attention to discourse and class has also been crucial in Andrew's important critique of labour and scale. Rather than adopting the dominant accounts of globalization as a story of capital, Andrew has shown that workers can mobilize class action in a way that shows the layers and connections between different social and political scales. He argues:

[The] global scale should not be considered to be solely the scale of capital – that is to say, it is not just capital that is responsible for shaping economic geographies at the global scale, but workers and their organizations are also important. ... [U]nions and other workers' organizations have been active in the international arena going back to the mid-nineteenth century. [Those of us on the left need] a political commitment that the triumphalist rhetoric of neo-liberal globalizers should not write workers out of the discourses surrounding issues of globalization.

By concentrating on different labour organizations, Andrew has documented the politics of class that operates via interconnected scales in a variety of contemporary settings (Herod, 1995, 1998b, 1998c, 2001b). Cumulatively, this work shows how geographies of class can adopt a core theoretical heritage (e.g. Marxism) and further engage with newer approaches (like poststructuralism) to show how a form of social difference such as class can form not a monolithic label, but an opportunity to mobilize economic and social interests through a range of labour actions and spatial interventions.

Jo Little is a Reader in Social and Rural Geography at the University of Exeter. She is one of many geographers who contributed to the growth in diverse and critical feminist geographies that have been written over the past 30 years. Just as feminist geographers first wished to highlight the heterogeneity and uneven power relations existing in everyday life and the topics conventionally studied in geography, so too does this geographer illustrate the heterogeneity and occasional marginality existing in geographies of gender. By devoting much of her work to questions of gender in rural societies, Jo continues to remind gender analysts that geography (and position/s) matters; that women's experiences will be uneven not only in terms of gender, class, race or sexuality, but also in terms of the broader social contexts and environments in which their lives are lived. This appendix records the key themes Jo has highlighted in this way. It also illustrates one geographer's experience of the varied political and theoretical trends that have constituted the dynamic growth of gendered geographies.

Jo completed an undergraduate degree in Geography at the University of Wales. She then moved to the University of Reading for her PhD work on rural geography and planning. Although her PhD was not specifically on gender, at Reading she was exposed to feminist perspectives as they were starting to be developed and used in human geography. She explains:

> The most important factor in the development of both my academic and political feminism was doing my PhD at Reading and being close to, if not a central part of, the very early establishment of the Institute of British Geographers' Gender and Geography group. Being in an environment where people like Sophie Bowlby, Linda McDowell, Linda Peak, Jane Lewis and Jo Foord were writing those first, crucial articles on feminist geography (including *Geography and Gender*, 1984) was very exciting and influential. Although obviously aware of the gender divisions in geography and the lack of women in senior positions, Reading was a little different and I felt more powerful there as a 'feminist geographer'. (Interview, 2002)

As noted in Chapter 4, much of the early focus of 'Geography and Gender' involved 'adding women' into the established sub-disciplines and topics of human geography. In Jo's case this activity was taken up from within her interests in both planning and rural geography. Works followed on feminist perspectives in rural geography (Little, 1986, 1987). Of this period, she reflects: 'My first sortie into writing on gender ... was very much about simply drawing attention to women's lives and experiences in rural communities. It was definitely adding women and stirring' (Interview, 2002).

After completing her PhD and working for a year as a postdoctoral researcher at Lampeter, Jo moved to University College London to work on a project on agricultural transformation and countryside change. During this time she worked alongside Sarah Whatmore and others, who were starting to apply feminist perspectives to the study of agriculture, looking, in particular, at the role of farm women (Whatmore et al., 1986, 1987a, 1987b, 1991).

While rural geography (and feminist critiques of it) had been dominated by agricultural studies, Jo Little drew on her broader planning and agency interests to document and highlight the breadth of other ways women experienced gender relations, inequalities and challenges in rural areas (Little, 1991a). Key work in this field included her account of women's non-agricultural employment, and women's experiences and positions that were negotiated beyond farm settings and within rural villages, communities and voluntary organizations (Little, 1991b, 1994a, 1997; Little and Austin, 1996).

> Having seen the development of feminist perspectives in the study of agriculture, I was keen to show how gender issues were relevant to rural people and places more broadly. I had enjoyed reading work on gender divisions by urban geographers and was very struck by the wider geographical applicability of issues such as access to services, transport and employment. As soon as I had written my first article on rural gender ('An Introduction to Feminist Rural Geography' – *Journal of Rural Studies*), which was really simply an account of rural women's roles, I was aware of the need to explore more theoretical debates concerning gender relations and the operation of patriarchy in a specifically rural context too. (Interview, 2002)

Intersections between gendered experiences and power relations have formed an ongoing core of Jo Little's work. Most recently this has continued in work focused comprehensively on gender (Halliday and Little, 2001; Little, 2002b) and that which continues her interests in planning and policy arenas (Little, 1999, 2001; Little and Jones, 2000). Alongside this core, Jo has also been engaged with newer thinking about gender and identity. Just as Chapter 4 recorded the increasing attention being given to **post-structuralist** readings of gender identities (including forms of masculinity),

Jo Little's work has also turned to consider the ways in which gendered identities can be theorized and read in rural contexts (Little, forthcoming; Little and Jones, 2000; Little and Leyshon, 2003). She describes the process of combining early critical theoretical perspectives on uneven and hegemonic power relations with newer poststructural and cultural interests in the construction and articulation of identities.

> My current work is trying to take the notions of gender identity and sexual identity further in a rural context. ... I think it is the development of work on masculinity and femininity and on the performance of sexual identity that has been very influential to my thinking. ... I have a strong sympathy with work on gender identity that has emerged from the cultural turn in geography but would be reluctant to let go of some of the ideas around the construction of hegemonic masculinities and femininities and on power relations inscribed through patriarchy that characterized earlier theoretical feminist work. (Interview, 2002)

This outlook enables Jo to continue a critical approach to gender issues while also using greater foci on identity performance and the embodiment and practice of sexualities that are taken up in poststructuralist and queer theory (see also sections 4.3.3 and 6.4).

Kay Anderson is a Professor of Geography at the University of Durham and her work in Australia and North America has influenced social and cultural geography because of the key accounts and questions she has presented on matters of race and ethnicity. Kay completed her first degree at the University of Adelaide in South Australia where she made contact with Professor Fay Gale, a 'charismatic teacher', whose work on 'Urban Aborigines' (Gale with Brookman, 1972) had a 'decisive influence' on Anderson's first understandings of race and ethnicity (Anderson and Jacobs, 1997). This experience led her to complete an Honours thesis on the refugee resettlement of Vietnamese 'boat people' in Adelaide.

The combination of her undergraduate education and her Honours research were crucial in leading Kay to start questioning ideas of race and difference, as well as the practice of conducting such research. She recalls:

> ... during the interviews ... with a sample of 'boat people' from Vietnam, I grew distinctly uneasy. ... I wondered whether anyone would ever consider knocking on the doors of the private homespaces of people like myself, a white settler of Scottish heritage, and asking them about the complex negotiations that make up everyday life! At the same time, the knowledge I had acquired about the experiences of Aboriginal people in white settler Australia, sharpened in me an equally acute sensitivity to the workings of power among majority groups. The impulse was ... one of ... a burning curiosity about that which gives powerful groups their self-appointed confidence and conceit. (Interview, 2002)

This interest was pursued at the University of British Columbia (UBC) in the mid-late 1980s when Kay completed her doctorate. She recognized preceding spatial projects that had 'mapped race' and developed explanations of ghettos and enclaves (as discussed in Chapter 5, section 5.3.3). However, the newly emerging cultural and discursive directions in geography at UBC were to stimulate her while she wrote her account of the multi-part construction of Chinatown in Vancouver:

The mid-late 1980s were lively ones at the graduate school at UBC Geography, where certain faculty were at the cutting edge of the development of geography's 'discursive turn' and cultural forms of analysis. So, while it might have been possible to write a thesis on Vancouver's Chinatown in the tradition of conventional enclave studies, the spirit of poststructuralist critique that was alive at UBC proved infectious for me, and I grew interested in alternative theorizations. Armed with Edward Said's (1978) work on Orientalism and an emergent critical race theory in sociology and anthropology, I 'wrote' Chinatown as a Western concept and material product. (Interview, 2002)

Kay's PhD thesis and the subsequent papers and book (e.g. Anderson, 1987, 1988, 1991) were important for demonstrating a genealogy of race through processes of institutional *racialization*. As discussed in Chapter 5 (section 5.3.3), Anderson has shown how racialization operates at individual, institutional, local and national levels, and with changing forms over time. Since then, with more reflection, and a re-engagement with the cross-cutting axes of difference formed by class, gender and race, she has re-interrogated her work to include more nuances:

On returning from Canada to Australia, and in view of key class alliances that were at work in structuring the redevelopment of Chinatowns in Sydney and Melbourne from the late 1970s onwards [1990, 1993b], I wrote an auto-critique [1998b] of 'Vancouver's Chinatown' from a political economy direction. Likewise, I chose to see what happened to the argument of 'Vancouver's Chinatown' when key moments in its history were recast through the lens of gendered vectors of power [Anderson, 1996]. (Interview, 2002)

This work illustrates the trend noted in the conclusion of Chapter 5, where geographies of race and ethnicity are pursuing ways to register more richly the multiple experiences and negotiations of social difference (and at times, discrimination).

The spatiality of these experiences and power relations has been further considered as Kay continues work on racialized minorities. In contrast to her migrant studies of Chinese in Canada and Australia, her research on Aboriginal settlement in central Sydney opened new perspectives on how 'race' can be constructed, and contested, from inner urban streets through to 'national debates' and post-national imaginings (e.g. Anderson, 1993a, 1999, 2000b). She explains:

[Redfern is the] site of Australia's first urban landrights struggle, existing within a few kilometres of the upscale heart of that city's central business district [and] has been the blighted home to Aboriginal tenants for some 30 years. For me, it opened a lens on to the experience of a very different racialized minority from that of immigrant groups such as the Chinese, with their consumption spaces called 'Chinatowns'. ... Taken together, the Sydney enclaves of Aboriginal Redfern and Chinatown prompted me to think more broadly about the diverse articulations of racialization in settler societies like Australia. ... I grew interested in the intersections of global migrant flows, and the relations of colonial capitalism, and in finding ways of thinking *across* the racialized categories of white, ethnic, and indigene. To that end, a 'postnational' perspective on pluralism and belonging has been fertile. (Interview, 2002)

While **poststructural** and **postcolonial** approaches have been important for invigorating geographies of race and ethnicity and showing the culturally and materially constructed character of these notions, Kay Anderson's work continues to question and disturb the power relations involved in these constructions. Most recently, she has been revisiting nineteenth-century British racial discourse in colonial Australia (Anderson, 2002, forthcoming). She states:

I have been seeking in various ways to disturb the universalizing claims of the 'civilizational discourse' through which Aboriginal Australia was constructed from the time of British 'settlement'. The Aboriginal savage (see Anderson, 1998a, 2000a, 2001) prompted an anxious crisis in the Enlightenment premise of human exceptionalism from the 1840s. He [sic] problematized the very notion of 'the human' as the being who transcends the (merely) natural. It is this broad intellectual context for the rise of nineteenth-century racial discourse that has most intrigued me. But rather than grasp this context of ideas as might a historian or a philosopher, I have sought local 'entry points' into it, such as the colonial performances of whiteness at Sydney's Royal Agricultural Show. These events open the intersections of culture, nature and colonialism. (Interview, 2002)

Work of this kind illustrates the way Kay Anderson continues to extend geographies of race and ethnicity, including of whiteness. Not only does she note the contexts, spatial relations and practices that constitute particular notions of race and ethnicity, but she challenges us to see links between these views and the wider cultures and philosophies in which they exist.

Larry Knopp: From 'mainstream radicalism' to explicit challenges
and contradictions in (sexualized) geography products and practice
Accompanying Chapter 6: Sexuality

Larry Knopp is a Professor of Geography at the University of Minnesota. Over the last two decades his work has paralleled and illustrated many of the trends and questions that have characterized geographies of sexuality. His work illustrates the 'contradictory' quality of sexual geographies and the ethical and political dimensions that infuse this type of work and the choices geographers make when conducting their work and academic careers (Knopp, 1999).

Larry developed an interest in geography after completing an undergraduate degree in Political Science. While studying for a Master of Arts (1986) and a PhD (1989) in Geography at the University of Iowa, he worked within an environment where radical Marxist geography was to some extent appreciated and given room to develop. As shown in Chapter 2, radical Marxist and feminist perspectives on geography were flourishing throughout the 1980s. From this early radical setting in his career, Larry was able to adopt and apply Marxist urban political economy perspectives together with early feminist and lesbian/gay thinking. Both individually, and jointly with Mickey Lauria, he was able to write important pieces on this work for the early gay/lesbian geography literature (Lauria and Knopp, 1985; Knopp and Lauria, 1987; Knopp, 1987). Looking back, he notes:

> My early work was strongly influenced by Marxist urban political economy and focused on urban land and housing markets, with issues of gender and sexuality 'added on'. It represented a fairly crude marriage of Marxist urban political economy, feminist, and lesbian/gay theories of politics and space. (Interview, 2002)

Although Larry's doctoral work and following publications concentrated on urban space and housing markets (Knopp, 1990a, 1990b, 1992), his interest in gender and sexuality was already established, although carefully positioned within the local department. He recalls:

> The University of Iowa Department of Geography in the 1980s was a pretty masculinist and patriarchal place. This was reflected in both the mainstream

and 'radical' perspectives represented there. My work was, in retrospect, adroitly positioned so as not to challenge this too much while at the same time appearing 'cutting edge'. (Interview, 2002; see also Knopp, 1999: 117)

While his early writing made important contributions to social and urban geographers' understanding of sexuality and power relations from a Marxist analytical perspective, Larry found the wider debates about social difference and the challenge of feminist geographers to be influential in the expansion of his work. He believes:

> International travel and critical intellectual challenges posed by colleagues – especial feminist colleagues – have forced me to re-evaluate my theoretical perspectives about class, gender and sexuality and the relationships among these social forces. ... Books like Gillian Rose's *Feminism and Geography* [1993] have caused me to completely re-think the discipline and my place in it as a male. ... This is good. I have grown tremendously as a result of these experiences and challenges. (Interview, 2002)

These trends and critical debates in geography influenced how Larry came to understand his experience of difference and his multiply constituted identity. He understood that certain attributes of class, gender and colour – and later occupational tenure – were more privileged than others. He recognized that he enjoyed some privileges and political freedoms while also facing ethical questions and implications for others (students and untenured staff) who have been allied with him (Knopp, 1999). Thus Larry's contemporary work, as a geographer committed to critiquing sexuality and academia (Knopp, 1994, 1997, 1999, 2000), is one that well illustrates the political and contradictory conditions of conducting geographies of sexuality. In terms of topics, his work has moved from the study of urban space *per se* to recognize and incorporate poststructural ideas about identity and power: 'My more recent work is much more influenced by postmodern and poststructuralist critiques of social science and social theory. It focuses on the spatiality of power rather than particular kinds of spaces as cities' (Interview, 2002). These themes are noted more broadly in the closing sections of Chapter 6 and taken up further in Chapters 7 and 8. In Larry's case, his reading of gay identities and cultures, the politics of institutional and discursive accounts of sexuality (Knopp, 1997, 1998) and his construction of queer critiques (Knopp and Brown, 2002; Knopp and Brown, forthcoming) each demonstrate these trends.

Larry Knopp's experience and work in geographies of sexuality again reinforce the dynamic quality of the politics and content of social geographies of difference. As with the previous biographies included in this book, this scholar has worked from within particular positions and engaged with the different broad theoretical currents moving throughout geography. As with those others, Larry Knopp's work and practice as a geographer demonstrates ways to name and negotiate the spatialization of difference and power as an ongoing contradictory and reflective process.

Karen M. Morin is an Associate Professor in Geography at Bucknell University, Lewisburg, Pennsylvania. During the last decade she has established herself as a feminist historical geographer with specific interests in travel, writing, and **postcolonial** issues associated with nineteenth-century women's travel. While this might appear to be a highly specialized interest in social geography, her work provides new angles and examples for understanding questions of identity (and gender) in social geography.

Karen initially completed studies in Philosophy and English and produced a Masters on writing for the environmental sciences before working as an editorial assistant for *The Great Plains Quarterly*. Through this experience she developed an interest in travel writing and texts that portrayed the American frontier and the imperial context in which this frontier was understood. Extending this interest during the early 1990s, Karen studied for a doctorate at the University of Nebraska, which had an established focus in historical geography. She recalls that from within 'a very positivist department of geographers' she chose to employ **poststructuralist** and **feminist** perspectives in her work. She drew on Foucault (1981) and Sara Mills's (1996) feminist, poststructural account of discourse and difference to present a gender-sensitive reading of how Victorian women travellers were depicting the American West (and themselves) in their own writings (Morin, 1995, 1996).

Of specific interest to our discussion of identity, Morin's work illustrates how a geographer can use wider radical and cultural theory to highlight how identities are formed and articulated – for both people and places. She comments:

> I position myself within a large, Anglophone, multi-disciplinary group of scholars working in what is now broadly called cultural studies. Much of the cultural studies scholarship (and its subdivision, postcolonial studies), including my own, is theoretically eclectic, drawing on feminist, poststructuralist, and Marxist theories and methods. (Interview, 2002)

This breadth of thought provides Karen with conceptual insights about identity, and the discourses and contexts from which they are constructed and shaped. Using **poststructural** theory, Karen understands identity as fluid and drawn from the multiple subject positions (or **subjectivities**) and social spaces that are discursively constructed through various texts. For instance, identities can be read from the diaries, travel logs and institutional regulations that people negotiate, as well as day-to-day decisions (e.g. dress) and social relations they encounter based on race, class and gender. Taking the writings of Victorian women travellers, she has shown how both the women – and the people and places they encountered – could be represented through a variety of narratives. These women's accounts of themselves, and the situations they experienced, provided texts that Morin (1998a, 1998b) read as geographically and socially varied:

> [E]ncounters between people of vastly unequal relations of power (for example nineteenth-century colonial administrators and Native Americans), and their resultant subjective senses of self, varied tremendously according to the location and social context of their encounters. Social relations on board moving trains, and at train stations, offered me particularly useful contexts for examining colonial subjectivity. (Interview, 2002)

While Morin's writings (e.g. 1995, 1999) show the different ways women identified themselves and gender relations more widely in the nineteenth century, she has also provided significant accounts of how identity is not only personally, but socially and politically, contextualized. In the case of her American research, and also her New Zealand collaboration (discussed in Chapter 7), Karen shows how identities associated with people and places represent a range of wider social discourses and power relations that frame our understandings about gender, race, politics and control. Thus, identities can be read not only for specific meanings about individuals and places (e.g. mountaineers and mountains – see section 7.3.3), but also for the masculinist, imperial and colonial beliefs and practices that prevailed in the USA and New Zealand in the nineteenth century (Morin, 1999, 2002, 2003; Morin et al., 2001). In this sense, Morin's approach to subjectivity and identity is intensely geographic and historic. She recognizes that specific places (and people's emplacement within them) can carry identities that are discursively constructed, but furthermore, she shows that geographers need to be attentive to the place (and the historic power relations and knowledge that exist). These place- and history-specific contexts colour and shape both the identities and relations between people: '... [G]eography is a main social axis of identity and difference ... including being an axis for understanding hybrid, national, complicit, and resistant identities (among others)' (Interview, 2002). For instance, using **postcolonial** readings of American imperialism, Karen is investigating how individuals (and groups) may be both simultaneously complicit with and resistant to colonial relations of power (Morin, 2002). Overall, Karen Morin's work shows how broader philosophies and theories of history and geography are intimately connected with how we understand and construct identities of ourselves, other people/groups and places.

Michael Woods: Developing an eclectic critical engagement with theories of power
Accompanying Chapter 8: Power

Michael Woods is a Lecturer in Geography at the University of Wales, Aberystwyth. During the last decade he has established himself as a geographer interested in the entwining of social and political questions. Growing up in a rural area enabled him to see that social and economic restructuring of the countryside resulted in local conflicts and changing power structures. This interest continued in the mid-1990s as he completed a doctorate on local rural politics and then developed more recent research programmes that have investigated community participation in rural politics, and the social and political character of rural activism in Britain, France and the USA. In each case, a consideration of notions of power has been central to his work.

Michael's work demonstrates the challenges and strategies geographers negotiate when faced with questions of power and politics. While recognizing established theories, they must also try out newer perspectives in both conceptual and empirical studies. For instance, he notes that his early work had been a reaction to the well-established structuralist approaches to power in urban geography and the political-economy analyses in rural studies that explained power relations in terms of class. In relation to the rural literature he studied for his PhD, he says:

> [In the 1970s] Newby [1977] had adopted a political-economy perspective and tended to approach power in terms of class. To me, Newby's work needed updating, both empirically and theoretically. I felt that individuals, and the networks of which they were part, were more important and I was drawn to elements of elite theory and, later, social network theory. (Interview, 2002)

Just as the context of geographers' work was shown to be important in Chapter 2, this was also the case for Michael. His doctoral work at the University of Bristol included interaction with 'staff and graduate students, many of whom were cultural geographers surfing the new waves of post-structuralist theory'. In this environment, Michael was able to question the degree of emphasis structuralist theories of power placed on institutions and economic processes. In contrast, he chose to concentrate more on geographies of power constructed from attention to **discourse** and people.

In the first instance, Michael's interest in Foucault's work led him to consider how power is contingent and articulated through discourse. He notes:

> I liked Foucault's emphasis on the contingency of power, and the way in which his notion of power seemed to be intrinsically geographical. I also felt that thinking about 'discourses of power' could help to explain how local power structures changed over time, and could be linked to narratives of rural change. (Interview, 2002)

Using a discursive approach to read the local geographies of power in rural Somerset, Woods' (1997) example of changing power structures showed how different narratives of rurality were linked to different interest groups who led local government over the twentieth century. This work formed one of the case studies in Chapter 8 (see section 8.3.1).

In the second instance, Michael's interest in power has turned to consider people's agency, both as influential individuals and as groups working in networks and movements. As a result of this decision, Michael has explored the use of elite theory and actor network theory to see how elites have worked in local rural politics (1998c) and how networks can be mobilized to engage with an issue (e.g. fox hunting: Woods 1998b). He recalls: 'Actor Network Theory particularly appealed to me because it revealed power as something that can only be exercised through the construction of a network of entities whose agency is combined to achieve a desired goal' (Interview, 2002). This construction and mobilization of power through networks and movements continues to interest Michael as he considers how people participate in local and national rural politics (Woods, 1998a; Woods et al., 2001). This work involves the writing of a 'new critical politics of rural citizenship' (Woods, forthcoming) which draws on:

> ... a range of theories and conceptual ideas that include theories of leadership and participation, citizenship and social movements, and governance and governmentality. Each of these approaches helps either to illuminate the context in which power relations operate, or to deconstruct the processes involved in the exercise of power. (Interview, 2002)

Using a combination of local politics and national rural protest events (Woods, 2003) and the theoretical debates circulating within geography, Michael demonstrates that one can establish an eclectic critical engagement with notions of power. He reflects that:

> I have tended to take an eclectic approach, borrowing from different theoretical perspectives rather than tying my colours to any one. This works so long as, firstly, care is taken to ensure that perspectives are only

drawn together where they are actually complementary not contradictory; and secondly, theories are critiqued and not just applied. (Interview, 2002)

Cumulatively, this widespread survey of applicable theories has resulted in four useful propositions in the study of power, each of which is touched on to varying degrees within Chapter 8, namely:

(1) power is a capacity not a property; (2) power is contingent and must be constructed, assembled by pooling different resources; (3) individuals, social networks and discourses are all important in facilitating power; [and] (4) the construction and exercise of power always has a geographical expression. (Interview, 2002)

J.K. Gibson-Graham: Creating geographies of
action – collaborative and alternative
Accompanying Chapter 9: Social Action

J.K. Gibson-Graham is the persona created by Katherine Gibson and Julie Graham during a gender study group meeting of the International Geographical Union in 1993. These economic and social geographers had worked together in various ways since they met in graduate school at Clark University in the mid-1970s. While studying at Clark, they submitted co-authored papers for several of their courses and have written together extensively. The 'birth' of J.K. Gibson-Graham 'honoured the third person our collaboration had become and cemented our collaborative affinity' (Interview, 2002). Interestingly, they found that it also liberated them to become more experimental and adventurous in their thinking and writing.

The choice of creating a joint identity and collaborative authorship is one example of the commitment Katherine and Julie have shared to building community and solidarity within the academy. While they are identified with the sub-discipline of economic geography, their work is a multifaceted resource for social geographers because of their longstanding interest in academic practices as political interventions. This brief biographical sketch can only touch on some of the projects they have completed, but it seeks to emphasize both the content and practice of their geography – as it informs our consideration of how people can, and do, act to create more socially connected and economically empowered communities.

Prior to their meeting, Julie Graham had completed an English degree and gone on to become an activist in the women's movement in Massachusetts where she built her own house and worked (in her spare time) as an editor and tutor. Katherine Gibson had completed a Geography degree in Australia, including an honours thesis in humanistic geography, looking at the 'sense of place' among long-term residents in an inner-city suburb of Sydney. Once at Clark, Julie and Katherine found a department/university environment energized by the radical possibilities of Marxist theory and the social movements that were proliferating in the USA and elsewhere. (This was the world of radical geography described in Chapter 2, especially section 2.3):

We were part of a first wave of geographers to be trained in Marxism during doctoral studies (in other words, we were students of the likes of David Harvey, Dick Peet, Ron Horvath, Bill Bunge and many others who had been trained in mainstream geography and who, as academics already teaching, retrained themselves in Marxism under the influence of the political movements of the 1960s and early 1970s – civil rights, anti-Vietnam War, etc.). The reading group we participated in as graduate students at Clark was very interested in the Althusserian notion of collective theoretical production as a form of political action. We saw ourselves generating new analyses that would inform political movements (primarily the labor movement), helping them to intervene in processes of internationalization (or what subsequently became known as globalization). We also saw ourselves as reframing what it was to be an academic producer – collectivity was valued and experiments in collaborative labor processes encouraged. Our decision to work together was never questioned. (Gibson-Graham, Interview, 2002)

Katherine and Julie's work was initially aimed at 'updating Marxian political economy, trying to theorize and track empirically the major changes taking place in the industrial, social and spatial organization of capitalist societies' (interview, 2002; and see Gibson et al., 1989; Graham et al., 1988). Their original concern was with industrial change, starting with the 1970s' wave of plant closings and the widespread de-industrialization that left workers and communities stranded in older industrial regions. This focus led them on to the embattled (masculine) terrain of Marxian crisis theory and long-wave theories of capitalist development, where they sought to understand the conditions promoting increased capital mobility and international production, and to trace the new forms of labour deployment that accompanied these world-changing shifts in capitalist practice.

Several events conspired to interrupt this early research programme, which had given Julie and Katherine the strong background in Marxian political economy that would inform their subsequent work:

- Julie encountered the Amherst school of *anti-essentialist* Marxists when she took up her job at the University of Massachusetts. Exposure to this school, which focuses on economy and class while refusing to accord these categories a privileged ontological status, allowed the duo to critique the *realist* and universalizing Marxist perspective they had first adopted (see Graham, 1990, 1992, and discussion of Gibson-Graham's perspective on class as *overdetermined* in Chapter 3, section 3.3). It also introduced them to a non-structural, non-systemic view of capitalism as a set of practices, and to the idea that capitalist and non-capitalist class relations can (and usually do) coexist in the economic landscape.

- Katherine embarked on a project of feminist action research in the coalfields of central Queensland (Gibson, 1992). This project provided the raw materials for Gibson-Graham's theory of 'poststructuralist action research' as involving the production of new discourses that could prompt novel identifications, engendering new subjectivities and possibilities of action (see Gibson-Graham, 1994).
- Soon after, Katherine took a job as Chair of Women's Studies at Monash University where **poststructuralist** feminist theory offered a way of questioning epistemological foundations and fixed identities, helping to shape Gibson-Graham's theoretical work on discourses of capitalism. Importantly, poststructuralist feminism introduced them to the performativity of discourse – the power of discourse to produce social effects – and reinforced their view that the production of new knowledges was a form of political action particularly suited to academic research.
- Julie started to explore queer theory, which went beyond binary difference (heterosexual/homosexual, normal/deviant) and opened the field of sexuality to radical heterogeneity; this body of theory inspired Gibson-Graham to deconstruct the capitalist/non-capitalist binary, re-theorizing the economic field as a site of proliferating differences.
- Julie and Katherine began to experience 'disenchantment with the conservative nature of leftist, union politics which seemed to be committed to working with capitalism to get a better deal for certain workers. We began to see that our earlier analysis and presumption of capitalist dominance was contributing to the hopelessness of the anti-capitalist left' (Gibson-Graham, interview, 2002). Their practice of left politics moved from the familiar project of ameliorating capitalism to the more uncommon one of fostering non-capitalist alternatives and desires, something they began to pursue through action research.

The confluence of all these events and inspirations led, in 1996, to the publication of *The End of Capitalism (as we knew it): A Feminist Critique of Political Economy*. This book was a critique of economic representation, attempting to destabilize images of capitalism

as necessarily dominant and naturally expansive, and as more resilient and powerful than other economic forms. Such images were part of what made a practice of non-capitalist economic development unimaginable on the left. We wanted to supplant or supplement the economic monism of conventional and leftist economic discourse with a discourse of economic diversity that could contribute to a politics of economic innovation. Our goal was to represent non-capitalist activities ... as prevalent and viable here and now, and thus as something we could build upon and replicate. (Gibson-Graham, interview, 2002)

Following the publication of *The End of Capitalism,* Gibson-Graham's economic research became a form of political action in itself:

> If knowledge is not assigned the task of providing an accurate reflection of reality ... then research doesn't simply reveal 'what's out there' in the world. Recognizing the effectivity of knowledge creates an important role for research as an activity of producing and transforming discourses, creating new subject positions and imaginative possibilities that can animate political projects and desires. (Gibson-Graham, 2000: 105)

Both Katherine and Julie undertook community-based research in regional settings, attempting to generate local knowledges of diverse regional economies. Drawing on anti-essentialist Marxian class theory as well as poststructural, queer and feminist theories, their research projects focused on both languages of economy and the embodied ***performance*** of alternative economic and social arrangements in people's everyday lives. In a process of academic–community collaborative action, they worked with community researchers in regions that had previously been seen as challenged or marginal in relation to a restructuring and globalizing capitalist world economy. The mainstream discourse of development positioned these regions as entirely dependent on investment by capitalist firms. This familiar vision had assumed the status of 'grim reality' and made it nearly impossible to talk hopefully about existing regional capacities and potentials:

> We used participatory action research to document and encourage alternative economic identities, organizations and actions. The action research projects produced alternative 'accountings' of the regional economy, ones that recognized a variety of non-capitalist economic relations and economic subjects. This alternative accounting served to destabilize the identity of our regions as failed or fragile sites of capitalist development, and was the first step in creating opportunities for new regional actors and different economic subjects to come together to talk about and construct alternative regional futures. (Gibson-Graham, interview, 2002)

Most recently, Gibson-Graham has focused on the intractability of existing subjectivities, especially as they are manifest in an attachment to capitalism and to the identities and fantasies it permits and promotes. Exploring wider notions of embodiment, performance, discipline and desire (Connolly, 1995; Foucault, 1997; Gibson-Graham, 2003), they have begun to theorize and practise 'a micro-politics of becoming, cultivating subjects who can desire and inhabit an alternative economic landscape' (interview, 2002; see also Community Economies Collective, 2001).

Gibson-Graham sum up their thoughts on action and academic practice as follows:

- Alternative geographies can destabilize existing economic discourses and representations to make room for novel kinds of knowledge and action.
- Alternative languages and representations can be generated through an inter*active* process of research.
- Alternative geographies can cultivate and support the subjects and actions that produce alternative economies. (Interview, 2002)

Glossary

Action (social action): For the purposes of this book, action is taken to include the social activity of people as individuals and groups. It is an under-theorized term since in the past geographers have variously concentrated on notions stretching from individual behaviour to group activism and phenomena such as *agency* and *social movements*. In this text, however, action is inclusive of individual acts and choices as well as social and political interaction which cumulatively encompass the way people interact with each other and the societies in which they live. Action is highlighted in order to focus on the dynamics of social life, showing how people's lives and worlds are shaped through choices and activities. Action therefore enables geographers to consider where and how people construct, understand and perform identity; how they negotiate and/or challenge power relations implicated in social differences; and how they strategically join together in temporary or longer-term projects that shape their experiences, environments and opportunities.

Actor: A term broadly used to refer to individuals or collective groups or organizations who are able to act in a variety of ways. In *actor network theory* actors (or actants) may be human, non-human or hybrid combinations (see Whatmore, 1999).

Actor network theory (ANT): A theoretical perspective originating from a sociology of science and technology (Callon, 1986; Latour, 1991; Law, 1994). Actor network theory has been adopted by geographers seeking to understand social relations and actions where power is not held in a structural or centralized form but is articulated through a network of 'heterogeneous association' made up of diverse human and non-human actants and the intermediaries that link them, such as texts, money and technological objects (Murdoch, 1997). The theory challenges 'modernist' distinctions between nature and culture. Networks operate as actants are connected (and hold each other in place) via various identities and capacities that are attributed to them (hence the popularity of ANT in geographical studies of governance, policy and planning, e.g. Bingham and Thrift, 2000; McGuirk, 2000; Murdoch and Marsden, 1995). Power and action are spread and spun across the network (sometimes over great distances) and decisions and results are produced as actors are enlisted into networks through a process of 'translation'. For further details see Murdoch (1997).

Agency (human action): Concept usually associated with people's autonomy or capacity to act either individually or collectively. See also *structure-agency*. Different philosophical approaches treat this concept in contrasting ways. Humanism has taken human agency as part of the central human subject whose consciousness, agency and creativity shape the world and understandings of the world (Tuan, 1976). In contrast, actor network theory includes humans and non-humans in ideas of agency where actants are able to act in a variety of networked and hybrid (e.g. human being and machines) forms so that agency is the result of networks of associating actors (Whatmore, 1999). Finally, poststructural theories and anti-humanist approaches critique the possibility of an individual, coherent or unified sense of human agency. Instead, arguments are made for the notion of *subjectivity* that recognizes multiple and conflicting subject positions within different (scientific, institutional and popular) discourses (Smith, P., 1988).

Anti-essentialist: An approach developed from *poststructuralist* philosophy and theory that questions other views of essential properties and definitions (for example, as found in Marxist ideas about capitalist relations, see Graham, 1992). In social and economic geography, Gibson-Graham (1996a) have criticized essentialist theories of capitalism and class, and argued that they are *overdetermined* rather than the product of fundamental essences of structures.

Anti-naturalism: An approach or view that contends that the social world cannot be understood or studied in the same way as the 'natural' world. (In contrast, see *naturalism*.)

Anti-realism: Belief that the world only exists through thought and that our ideas or constructions are what constitute reality. This perspective complements *idealism* but is the opposite of *realism*.

Capital accumulation: A Marxist term for the central tendency by which capitalist systems seek to secure conditions and processes that will constantly enable the increasing growth of capital via the reinvestment of surplus value (see Harvey, 1977). Processes of economic growth and technological change regularly follow on from this tendency, however (unequal and spatially varying) conditions change under which capital can be accumulated and crises of accumulation also frequently occur in the form of recessions.

Chicago School: The name for the scholars and research produced in the early twentieth century through the Department of Sociology at the University of Chicago. This school of thought used *pragmatism* and ethnographic methods (especially participant observation) to study the operation of social groups within different areas in Chicago. This work has been particularly relevant to urban geographers seeking to understand the spatial structure and social processes that produce different city formations. The Chicago School concentrated on elements of social Darwinism and human ecology to study society using ecological metaphors and principles to understand the formation of communities and the invasion and succession of different groups within a city (Park et al., 1925). While the Chicago School has been widely criticized for its universalizing, *metanarrative* and *naturalist* qualities, Jackson (1984) argues that it has formed a key heritage for geographers choosing *humanist* and ethnographic approaches to social geography.

Class: A concept used to indicate social difference between groups of people based on their material circumstances and positions within the dominant economic system operating in their society. Within capitalist societies class is often expressed in Marxist analyses according to people's positions within the capitalist *mode of production* and the various labour relations that exist during different eras. Contrasting definitional attributes and theories of class are detailed with relevant references in Chapter 3.

Deductive: A way in which a general theory is constructed, after which data on particular cases and situations are gathered to test the validity of the theory. (In contrast, see *inductive*.) Deductive approaches are frequently thought of as 'top-down' thinking where theory is developed conceptually before it is tested through specific cases or field data.

Discourse, discursive: In this book, discourse is used to refer to the *poststructuralist* concept of how meanings, ideas and practices are assembled, circulated and per-formed as a way to understand the world or legitimize certain types of knowledge. Discourses are not single, unitary products of individual authors but are broad set of meanings and practices that are situated and embedded in particular contexts, societies and histories (Barrett, 1992; Weedon, 1987). They are performed and they circulate or resonate through varying institutions (e.g. law) and day-to-day practices (e.g. conversations). Geographers look at how broad social relations and differences like gender and ethnicity are constructed through discourses. They also reflect on how their own discipline is a complex product of multiple discourses (e.g. colo-nialist and scientific, see Crush, 1991; Driver, 1992).

Epistemology: The theory of knowledge, how it is derived and how it can be known (in contrast to *ontology*). Within human geography epistemologies underpin-ning *spatial science* have respectively valued scientific, objective, constructions of space and society. In turn these have been criticized (e.g. by *humanist* and *feminist* epistemologies) for overvaluing scientific explanation and for being based on cer-tain privileged and dominant forms of western and masculine thought (see Entrikin, 1976; Rose, G., 1993).

Ethics: The study of morality and the evaluative judgements associated with what is moral. In geography, ethics have influenced both what geographers study (e.g. interests in inequalities, *marginalization*, exclusion and social justice) and how they conduct their geographies. During the rise of radical geography in the 1970s, key geographic works that took up ethical positions included Harvey's (1973) and D.M. Smith's (1977). More recently, geographers have increasingly considered the ethics underlying how they design and conduct their research as well as the value and purpose of that work (Cloke, 2002; Hay, 1998) and have established the journal *Ethics, Place and Environment* in order to pursue these issues further.

Ethnicity: A concept used to indicate social difference between individuals and groups of people based on their identification with shared ancestry, culture, traditions, beliefs, and behaviour. Within western societies ethnicity is frequently understood as 'other' to the dominant ethnicity, i.e. *whiteness*, although recent research has begun to interrogate the ways whiteness is constructed and maintained as a dominant ethnicity. Discussion of the definitions and analyses of ethnicity that have been conducted by geographers are included with relevant references in Chapter 5.

Existentialism: A philosophy of science that views reality as existence that is prior to, or existing before, human essence. Consequently, focus is given to how individuals relate to other phenomena and make them meaningful. Existentialism is critical of both rationalism and *idealism* but has connections and contrasts with *phenomenology*. Existentialism has supported humanist geographers' interest in the geographies of people's everyday life and the meanings they have ascribed to a 'taken-for-granted world' (Buttimer, 1976, 1979; Samuels, 1978).

Feminist, feminism: A view of society and knowledge that suggests men's views and interests dominate and that unequal gender relations (both in society and research) disadvantage women (Bowlby et al., 1989; Mackenzie and Rose, 1983). Different forms of feminism have produced different types of explanation (see Blunt and Wills, 2000: Chapter 4; Tong, 1992) but irrespective of these differences, feminism aims to challenge and provide alternatives to existing structures, institutions, social relations and ways of conducting geography (Monk and Hanson, 1982; Rose, G., 1993).

Gender: A concept used to indicate social difference between individuals and groups of people based on socially constructed ideas that are attributed to men and women and understood in terms of masculinity and femininity. In geography, gender differences have been recorded since the 1970s, after which critical theoretical analyses were made to explain the frequently disadvantaged experiences of women. Since then, geographers have employed understandings of *gender relations* (in particular, *patriarchy*) to investigate uneven power relations based on expectations, rights and opportunities that men and women negotiate according to prevailing ideas about each gender. More recently, poststructural approaches have problematized ideas about an essential structural explanation of gender differences, pointing instead to the way masculinity and femininity are constructed notions that have both dominant and many other forms. Attention to *gender identities* has been a product of this latter thinking. Discussion of the definitions and theoretical approaches surrounding gender are included with relevant references in Chapter 4.

Gender identities: A concept most common in poststructural analyses of gender and one which focuses on how gender is understood and negotiated via broad social meanings of masculinity and femininity that are constructed and performed with only limited choice for men and women (Butler, 1993; Connell, 1995). These identities are understood as multiple, contested, and spatially and culturally situated so that while some meanings and practices surrounding certain forms of masculinity and femininity may dominate, there is also an appreciation of alternative forms that can be adopted, mobilized and performed (Jackson, 1991, 1999; Laurie et al., 1999; Little, 2002b).

Gender relations: A concept employed by both socialist and radical feminists in order to recognize and analyse the specific power relations that are based on, and affect peoples' experiences of, gender. Gender relations may take numerous forms but the most widely acknowledged and researched is that of *patriarchy* (see, for example, Foord and Gregson, 1986; Knopp and Lauria, 1987).

Hermeneutic: A form of creating knowledge and conducting research that is focused on interpretations and meanings. These interpretations are drawn from the study of social texts, symbols and meanings that people ascribe to everyday life. Within geography, *humanist* approaches employed hermeneutics to document and interpret human meaning and to read texts for the meanings they convey about people and the places and spaces they inhabit and shape.

Heteronormative: The widespread and dominating assumption that heterosexuality is both the natural/normal form of sexuality and the most appropriate or best form. Geographers investigating heterosexuality and using *queer* theory have shown how heteronormative beliefs and practices can be experienced through academic research, everyday speech, social expectations, control of public space, popular culture, and policies of various institutions and agencies (Binnie and Valentine, 1999; Grant, 2000; Hubbard, 2000).

Heteropatriarchal: A concept used to show how power relations based on *patriarchy* and heterosexuality can be combined to create dominant (and at times oppressive) control and expectations associated with social interaction and everyday places (Valentine, 1993a, 1993b).

Historical materialism: A term given to the approach employed by Marx and Engels to analyse the material (and historically dynamic) organization of society. This approach, and its practice by Marxist geographers (e.g. Harvey, 1984), counters *idealism* and prioritizes the material basis of life (over consciousness or ideas by which we might experience life). Historical materialism concentrates on establishing analyses and explanations of the social and economic processes by which people operate (and shape space) in various *modes of production*. In recent years it has been critiqued by *postmodern* and *poststructural* theorists for being a *metanarrative.*

Humanism, humanist: An approach to inquiry and knowledge that focuses of human awareness, values and creativity and in some cases concentrates on social meanings and actions generated by people as a product of essential attributes and *agency* that are thought to be common to human beings. Humanist geography is often grounded in *phenomenology* and employs qualitative and *hermeneutic* approaches (Buttimer, 1976, 1979; Tuan, 1976).

Hypothetic-deductive inquiry: An approach to social science that draws on *deductive* reasoning and establishes hypotheses that are tested in order to build explanation. As noted in Chapter 2, Harvey (1969) argued the importance of hypothetic-deductive inquiry for human geography and this complemented much of the quantitative and *spatial science* activity occurring throughout geography at the time.

Idealism: A philosophical position that contends that reality is formed through ideas or thoughts rather than predating them. This approach is often positioned in contrast to *realism*.

Identity: A term associated with the understanding of self or other people and places. Using poststructural approaches, geographers have investigated the way individual, collective and place-based identities are multiple (i.e. we have no single identity) and are constructed through discourses and performances (Hall, 1996; Jackson, 1999). This active or dynamic view of identity means that the contested and unstable nature of identities can be recognized. So, too, geographers have come to appreciate identity as a social and political process whereby understandings of self and other are explored (Hetherington, 2000) and the mobilization of strategic identities is possible (see *identity politics*). A detailed discussion of definitions and relevant references is included in Chapter 7 (section 7.2).

Identity politics: A term used to understand how social and cultural identities can be strategically assembled and enacted in politicized processes for collective goals. The study of identity politics became popular in geography in the 1990s (Keith and Pile, 1993) as cultural geography and poststructural theories stimulated new research into social movements that were committed to achieving anti-racist, *feminist* and *queer* goals. While poststructural thought had highlighted the socially constructed character of categories of social difference, identity politics allowed geographers to investigate how social difference was negotiated and contested in terms of politicized identities and socio-spatial activism associated with various forms of resistance (Pile and Keith, 1997).

Inductive: A way in which a theory or explanation is constructed from (or following after) the analysis of gathered data (in contrast see *deductive*). Inductive approaches are frequently thought of as 'bottom-up' thinking where theory is developed out of case material and field data.

Marae: 'Traditional Māori gathering place' (Reed, 2001: 42). Both traditional and newer urban *marae* are key material sites and cultural symbols of individual tribal groups' identities. They include communal territory, formal open-air space and the traditionally carved meeting house (*whare whakairo*).

Marginalization: A response to social difference which can be made at personal and institutional levels whereby certain individuals and groups are physically, socially, economically and/or culturally placed as marginal to the 'centre' where people and values are associated with dominant categories of social difference (e.g. masculine, heterosexual, white). Marginalization is not just about a metaphoric location. It is also associated with the social processes of othering and the creation of *otherness* – where difference is expressed as other to the self or the dominant social categories. Geographers' interest in marginalization and otherness has resonated in many critiques of prejudice and inequalities based on class, gender, ethnicity and sexuality, and it is frequently inspired by alternative politics and the opportunities of action from the margins (hooks, 1990, 1992; Spivak, 1990). Reflections on how marginalization operates within geography have also been published by authors who have identified how different types of academic (and research subject) have been excluded from geography (e.g. Monk and Hanson, 1982; Rose, G., 1993; Sibley, 1995).

Marxism, Marxist: A critical approach to science and knowledge based on the work of Karl Marx and employing ideas of *historical materialism*. This approach attempts to critique and change society by understanding the processes and relations that structure the material and social arrangements of different *modes of production*. Marxist geographers' particularly concentrate on documenting and challenging the injustices that result from unequal social and economic relations that exist in capitalism.

Masculinities: A range of identities that are constructed to convey understandings of what it is to be masculine. As shown in section 4.3.3, numerous geographers using poststructural notions of *identity* have established analyses of how masculinities are constructed and performed, noting that some forms dominate (in forms that can provoke masculinist assumptions) (Rose, G., 1993).

Metanarrative: A postmodern concept for a powerful and widely recognized frame of understanding usually derived from one perspective or emphasizing a particular principle that is privileged (or universalized) as a claim to truth, e.g. see *historical materialism*. It is a particular macro-view of the world that privileges certain 'truths' while marginalizing others.

Mode of production: A marxist concept for the way in which societies organize the material structures and activities that sustain life and extract surplus value. Different modes of production, such as slavery, feudalism and capitalism, have involved different economic and social relations by which societies operate unevenly in producing material conditions and reproducing social relations that maintain that society. Geographers have most frequently studied the social and economic relations associated with capitalist modes of production (Harvey, 1982, 1989).

Naturalism, naturalist: An approach or view that contends that the social world can be investigated using similar methods to those employed in the study of the 'natural' world Bhaskar (1979). Approaches and concepts used in biology and physics have been particularly popular for investigating society at different times. In contrast, see *anti-naturalism*.

Non-representational theory: A theory that seeks to challenge and balance the recent intense emphasis that has been given to geographies of interpretation and representation of a socially and culturally constructed world. The term has been introduced by Thrift (1996, 1997) to convey a move away from representation and *discursive* considerations so that work can be undertaken on the practice, *performance* and flow of events and transitory moments and interactions of everyday life and the actions by which it is constituted.

Ontology: A theory of what exists, what the world is like, or what can be known (in contrast to *epistemology*). Scientific inquiry is underscored by ontologies of how the world must be so that decisions can be made about what will be investigated. Contrasting ontologies focus on forms of empiricism, *idealism* and *realism* (Bhaskar, 1978) and these in turn affect how space and society are conceived even before research is designed and conducted (Schatzki, 1991).

Other, otherness: A process by which difference is interpreted through ideas and actions that differentiate between a binary of self and other. The construction of otherness is frequently associated with unequal power relations and processes of exclusion and *marginalization*. Geographers interested in many forms of difference have employed this idea to see how otherness is constructed through a range of categories and processes (e.g. those that are sexist, racialized, colonialist, abelist and heternormative).

Overdetermined: A term used by scholars working with anti-essentialist approaches when referring to how a social phenomenon or process (such as class) is mutually constituted and intersecting with other social phenomena and processes. Gibson-Graham's (1996a) discussion of class as an overdetermined category and process is one good example of this thinking (see Chapter 3, section 3.3.3).

Patriarchy: A particular form of gendered power relations by which men have more control and advantages than women. This uneven gender relation operates through economic, social and political practices and structures, including what Walby (1990) identifies as six spheres: the state, cultural institutions, wage-based employment, violence, sexuality, and the household. In recent years, feminists using patriarchy-based explanations of gender have also come to accept that other forms of *marginalization* and oppression also need to be considered (as noted in section 4.4).

Performance, performativity: A sense in which daily life, social differences, identity and power are enacted through practice or performance that both draws on social meanings and the spaces and bodies we inhabit (Butler, 1993; Hetherington, 1998). Geographers have appreciated ideas of performativity for the opportunity to see society and identity as dynamic and practised (even political) rather than only *discursive* or textual (Nash, 2000; Rose and Thrift, 2000).

Phenomenology: A philosophy of science that focuses on understanding (rather than explaining) the world. Reality is understood through the subjective meanings that people develop. This approach is often employed in *humanist* geography and is positioned in contrast to *existentialism* and *positivism* (Pickles, 1985; Relph, 1970).

Political subjectivity: A concept used in poststructural geographies focusing on politics and social action (e.g. Gibson-Graham, 1995). Drawing on notions of subjectivity as discursively produced in multiple forms, political subjectivity is seen to be mobilized when individuals or groups strategically negotiate contradictions or opportunities that exist between contrasting subject positions and thus provide themselves with a range of responses or actions that can be taken up (Davies, 1991). Examples of this type of analysis are discussed in the closing part of section 9.2.

Position, postionality: An acknowledgement that our location or position within a society will affect how we understand and value the world and create knowledge (Hartsock, 1987). While this notion has been most widely adopted within *feminism*, geographers have employed the idea to encourage critique of dominant social categories such as heterosexuality and *whiteness*. Consideration of positionality also stimulates geographers to consider the situated nature of geographic knowledge and the partial and positioned choices geographers make in the practice and construction of their geographies (Rose, 1997).

Positivism, positivist: A philosophy of science that employs empiric observations of positive phenomena (those which can be observed with the senses) using systematic procedures to test hypotheses in controlled ways and analyse results in order to construct formal laws and/or predictions (Guelke, 1978).

Postcolonialism, postcolonialist: A movement stemming from literary and cultural studies that has critiqued the processes and implications of historic and ongoing (reconfigured) colonialization for both colonized and colonizing and societies (Childs and Williams, 1997). Postcolonialism seeks to the contrasting knowledges, goals and positions of colonized people tive also supports questions concerning the dominance and auth Jacobs, 1996; characteristic of western thought and social structures as historic and ongoing processes and spaces by which colo colonized as *other* (Driver and Gilbert, 1998; Jackson and Mills, 1996).

Postmodernism, postmodernist: A move which critiques modernism and the pos suggested in various (often dominant) of heterogeneity and difference, po (after modernism) and a method recognizes the researcher as the multiple voices of modernity has been social and spatial Harvey, 1989;

Poststructuralist: An approach to knowledge and inquiry by concentrating on how society and the world is rally specific language and discourse (both the meanings ch we understand the world). In contrast to *humanism*, post-give attention to how people (individuals and groups) are constructed as subjects in different discourses that can be decon-multiple meanings and possibilities (see section 2.3).

A concept widely understood as a capacity to act or achieve desired effects. ography, power has been noted as present in economic and social relations as ll as cultural beliefs. Contrasting theories of power with relevant references are luded in Chapter 8. Main differences occur between conceptualizations of wer as a resource or capacity that can be held by some people and institutions d those that conceive power as multiply constituted and fluid phenomena that culate through networks and relations. The latter view sees power not so much ssessed as articulated through a myriad of practices and technologies that repro-ce it at the same time as providing points and opportunities for resistance. While itings on power have been a primary focus in much social geography, Nigel Thrift's 97, 2000) attention to 'dance' provides some challenges and alternatives to the sic argument being introduced in this book concerning power, especially when considers creativity and play, as he writes 'against the grain' to argue that power not 'all-pervasive'.

Pragmatism: A philosophy of science that focuses on understanding the world through real-life problems and the recognition of behaviour and the experience of practical daily life rather than knowledge we might construct about it.

Queer: A movement drawing on the twin meanings of lesbian/gay sexuality and strangeness or oddness in order to build a theoretical and political challenge to the *heteronormative* qualities of most accounts and theories of social life. Queer theory generally rejects notions of fixed or stable sexuality (or theory), drawing instead on the opportunities provided in language and ambiguity to challenge and unravel dominant forms of knowledge and sexual relations. For geographers, this includes challenging the heteronormative tendencies to fix sexual patterns and relations in stable spaces (e.g. gay ghettos or red-light districts), thus enabling a wider reading of sexuality as constantly played out through all social formations of space (Bell and Valentine, 1995a; Costello and Hodge, 1999).

Race: A concept used to indicate social difference between individuals and groups of people based on physical characteristics (see Chapter 5). While racial classifi-cations have since been repudiated as politically and socially produced, their early 'scientific' claims have been important for creating widespread prejudicial knowl-edge about the human capacities and values of different peoples during colonial periods. Geography was complicit in the construction of racialized discourse (Maddrell, 1998) but more recently has turned to investigate the social, cultural and spatial processes by which race is constructed (Bonnett, 1996; Jackson and Penrose, 1994). The racist impact of race-based thinking continues in the inertia of *racialization* that exists within major institutions and popular discourse.

Racialize, racialization: A process by which phenomena (e.g. people, places, discourse, behaviours, opportunities) are classified and controlled according to race-based beliefs. Racialization can affect people's access to resources (e.g. land, housing, citizenship) and understandings of space and place through a variety of means (e.g. urban districts, place names, national identities). Detailed examples of references to racialization are given in Chapter 5, especially section 5.3.3.

Realism, realist: A philosophical position that argues that the real world exists in material form beyond our ideas or conceptions of it (in contrast, see *anti-realism* and *idealism*). Realism is also recognized as a philosophy of science that is often posi-tioned in opposition with *positivism* since it values conceptual abstraction as a way to explain relations and causes between phenomena in specific cases.

Reflexive: The practices of being self-aware concerning the contexts affecting our knowledge and choices, and of questioning how our *positionality* and experiences affect our understandings and decisions while conducting research. Geographers adopting a view of knowledge as situated and partial most frequently use reflexive strategies to consider explicitly their positions and choices in relation to the geographies they are creating (Rose, 1997).

Resistance: Activities (from conceptual thinking to political action) that are committed to questioning, challenging or overthrowing existing power structures, relations and conditions. Geographers have used *resource mobilization theory, feminist theory, identity politics, social movement* studies and Foucauldian perspectives on power to investigate the ways people form resistant action and strategically use space in the process. Discussion of resistance is included in Chapter 8, section 8.2.2.

Resource mobilization theory (RMT): A theory that explains the capacity of people to mobilize power and achieve social action by assembling and using symbolic and material resources that they can access (to varying degrees) (see, for example, McAdam et al., 1996; Oberschall, 1973). RMT has not been widely adopted in social geography since most studies of social action have been completed from critical social perspectives (e.g. *Marxism* and *feminism*) while RMT is seen to be grounded in individualist and liberal models of public and political action.

Science: A form of knowledge and inquiry that is frequently based on *realist* or *positivist* philosophies and *ontologies*. Social sciences vary in terms of ontologies and *epistemology* but aim to produce knowledge (including some forms of human geography) through methods of design and measurement that purport to be systematic, objective, logical and rational. This perspective supported the development of geography as a *spatial science*. However, geographers have also critiqued the uneasy relationship geography has had with science, including Livingstone's (1992) argument that geography has been 'the science of imperialism'.

Sex: A problematic concept used to distinguish individuals according to biological and reproductive attributes. Early distinction was made between 'sex' as biologically determined and '*gender*' as socially constructed. However, more recent theorization of sex has contended that sex is also constructed, understood – and even managed – via powerful social and cultural frames of meaning (Butler, 1993; Pratt, 2000a).

Sexuality: A concept used to distinguish social difference based on sexual identity and behaviour. Geographers recognize this as a heterogeneous category even while acknowledging that *heteronormative* attitudes and processes predominate in the organization of society and space. The range of patterns, relations and politics surrounding sexuality are detailed with relevant references in Chapter 6.

Social movements: The formations of individuals and groups into strategic collectives that work for social and political goals using a range of strategies and spaces. Social movements have been investigated using a range of theories including *resource mobilization theory* and Melucci's (1996) conceptualization of collective action. New social movements are being recognized as a special form of collective action in late modern and postmodern societies (Crook et al., 1992).

Spatial science: A form of geography emerging from the wider quantitative revolution of the 1960s and 1970s. This geography is characterized by analysis of spatial systems and structures, including detailed statistical analyses, geometric theories and mathematical modelling. Proponents sought to produce generic understandings and robust predictions based on these quantitative methods and scientific approaches (e.g. Berry and Marble, 1968; Chorley and Haggett, 1967; Haggett, 1965).

Structuration theory: A theory developed by sociologist Anthony Giddens (1984) to explain the ongoing relations and processes linking individual action and agency with broader social structures. He retained an interest in human agency while also developing a theory that was attractive to those working from *realist* perspectives and who wished to be able to develop an abstract theoretical framework to understand and investigate relations and processes mutually linking people and social systems through a range of structures, resources and rules.

Structure-agency: A pair of concepts that are often linked to highlight questions about individuals' and groups' ability to act (*agency*) and the constraints or contexts (structure) that influence them. Some theorists (e.g. Marxists) focus on structure, considering social action to be predominantly affected by wider social structures. Other theorists (e.g. Behaviouralists) focus on agency, considering that individuals are autonomous and free to choose how they behave. Gidden's *structuration theory* (1984) sought to build an analytical framework that gave equal emphasis to each.

Subject, subjectivity: A concept most frequently used in poststructural theories to refer to the variety of subject positions that are formed through different discourses (Smith, P., 1988). In contrast to *humanist* beliefs in an essential human essence and unified *agency*, notions of subjectivity encourage recognition of the multiple (even hybrid), and frequently potentially contested possibilities available to human subjects (Pile and Thrift, 1995; Whatmore, 1999).

Symbolic interactionism: A social theory that identifies the social world as a project of social interaction. *Humanist* geographies have drawn on this perspective when analysing landscapes, places and environments as a socially constructed reality that reflects the social interaction and meanings human beings bring together in different cases/locations (Duncan, 1978; Ley, 1981).

Tangihanga: 'Funeral, wake' (Reed, 2001: 74). A key traditional cultural institution involving Maori customs associated with bringing a deceased person back to their *marae* for a period of mourning and burial. Customs vary depending on the *iwi* or tribe.

Tapu: 'Forbidden; inaccessible; not to be defiled; sacred; under restriction' (Reed, 2001: 74). A key concept by which Maori identify that which is sacred and deserving of respect or special ritual.

Theory: An explanation or framework that is devised and argued as a way to understand or interpret phenomena (social or otherwise). In social sciences such as social geography, theories are influenced by the type of philosophical approach adopted. Theories are built up using concepts to represent social states, conditions and processes (e.g. class, exploitation and labour relations). Depending on the type of theory, these concepts are linked together using propositions (e.g. Marxist theory contends that capitalist class structure produces exploitative conditions as a result of unequal labour relations). Most theories are debated and tested (formally or informally) through both conceptual debate and empirical research. For general overviews and discussion of theory in geography, see Hubbard et al. (2002: Chapter 1); Kitchin and Tate (2000).

Whanau: 'Family (in a broad sense)' (Reed, 2001: 98). A concept for Maori which refers not only to immediate and extended family but also to the extended group of people who can provide an individual with support and cultural connection.

Whiteness: A culturally constructed category of ethnicity that forms a base for the dominant or mainstream social knowledge, cultural expression and politics, against which alternative ethnicities, cultures, histories and frames of knowledge are compared as *other* and often *marginalized*. Geographies of ethnicity and race have come to interrogate the way power relations and cultural assumptions associated with whiteness have underpinned the organization of societies and the writing of geography (Bonnett, 1996, 1997; Jackson, 1998).

References

Adler, S. and Brenner, J. (1992) 'Gender and space: lesbians and gay men in the city', *International Journal of Urban and Regional Research*, 16: 24–34.

Agg, J. and Phillips, M. (1998) 'Neglected gender dimensions of rural social restructuring', in Boyle, P. and Halfacree, K. (eds), *Migration Issues in Rural Areas*. London: Wiley. pp. 252–79.

Agyeman, J. and Spooner, R. (1997) 'Ethnicity and the rural environment', in Cloke, P. and Little, J. (eds), *Contested Countryside: Otherness, Marginalisation and Rurality*. London: Routledge. pp. 197–217.

Aitken, S.C. (2001) 'Fielding diversity and moral integrity', *Ethics, Place & Environment*, 2: 125–9.

Akatiff, C. (1974) 'The march on the pentagon', *Annals of the Association of American Geographers*, 64: 26–33.

Allen, J. (1997) 'Economies of power and space', in Lee, R. and Willis, J. (eds), *Geographies of Economies*. London: Arnold. pp. 59–70.

Amin, A. and Thrift, N. (1992) 'Neo-Marshallian nodes in global networks', *International Journal of Urban and Regional Research*, 16: 571–87.

Anderson, K. (1987) 'The idea of Chinatown: the power of place and institutional practice in the making of a racial category', *Annals Association of American Geographers*, 77: 580–98.

Anderson, K. (1988) 'Cultural hegemony and the race definition process in Vancouver's Chinatown: 1880–1980', *Environment and Planning D: Society and Space*, 6: 127–49.

Anderson, K. (1990) 'Chinatown re-oriented: a critical analysis of redevelopment schemes in a Melbourne and Sydney enclave', *Australian Geographical Studies*, 28: 137–54.

Anderson, K. (1991) *Vancouver's Chinatown: Racial Discourse in Canada, 1875–1980*. Montreal and Buffalo, NY: McGill, Queen's University Press.

Anderson, K. (1993a) 'Place narratives and the origins of inner Sydney's aboriginal settlement, 1972–73', *Journal of Historical Geography*, 19: 314–35.

Anderson, K. (1993b) 'Otherness, culture and capital: Chinatown's transformation under Australian multiculturalism', in Clark, G., Francis, R. and Forbes, D. (eds), *Multiculturalism: Postmodernism, Image and Text*. Melbourne: Longman Cheshire. pp. 58–74.

Anderson, K. (1996) 'Engendering race research', in Duncan, N. (ed.), *Bodyspace: Destabilising Geographies of Gender and Sexuality*. London: Routledge. pp. 197–211.

Anderson, K. (1998a) 'Science and the savage: The Linnean Society of New South Wales, 1874–1900', *Ecumene: International Journal of Culture, Environment and Meaning*, 5: 125–43.

Anderson, K. (1998b) 'Sites of difference: beyond a cultural politics of race polarity', in Fincher, R. and Jacobs, J.M. (eds), *Cities of Difference*. New York: Guilford. pp. 201–25.

Anderson, K. (1999) 'Reflections on Redfern', in E. Stratford (ed.), *Australian Cultural Geography: A Reader*. Melbourne: Oxford University Press. pp. 67–84.

Anderson, K. (2000a) '"The beast within": race, humanity and animality', *Environment and Planning D: Society and Space*, 18: 301–20.

Anderson, K. (2000b) 'Thinking post-nationally: dialogue across multicultural, indigenous, and settler spaces', *Annals of the Association of American Geographers*, 90: 381–91.

Anderson, K. (2001) 'The nature of race', in Castree, N. and Braun, B. (eds), *Social Nature*. Oxford: Blackwells. pp. 64–83.

Anderson, K. (2002) 'Post-human geographies', in Anderson, K., Domosh, M., Pile, S. and Thrift, N. (eds), *Handbook of Cultural Geography*. London: Sage. pp. 18–23.

Anderson, K. (forthcoming) 'Race, culture, nature: Sydney's Agricultural Show in post-humanist perspective', *Transactions, Institute of British Geographers*.

Anderson, K. and Jacobs, J. (1997) 'From urban aborigines to aboriginality and the city: one path through the history of Australian cultural geography', *Australian Geographical Studies*, 35: 12–22.

Anderson, K. and Smith, S.J. (2001) 'Editorial: emotional geographies', *Transactions of the Institute of British Geographers*, 26: 7–10.

Barnes, T.J. (2001) 'Lives lived and lives told: biographies of geography's quantitative revolution', *Environment and Planning D: Society and Space*, 19: 409–29.

Barnes, T.J. (2002) 'Performing economic geography: two men, two books, and a cast of thousands', *Environment and Planning A*, 34: 487–512.

Barrett, M. (1992) *The Politics of Truth: From Marx to Foucault*. Stanford, CA: Stanford University Press.

Bell, D. (1994) 'Bi-sexuality – a place on the margins', in Whittle, S. (ed.), *The Margins of the City*. Aldershot: Ashgate. pp. 129–41.

Bell, D. (1995) 'Perverse dynamics, sexual citizenship and the transformation of intimacy', in Bell, D. and Valentine, G. (eds), *Mapping Desire: Geographies of Sexualities*. London and New York: Routledge. pp. 304–17.

Bell, D. (2000) 'Farm boys and wild men: rurality, masculinity and homosexuality', *Rural Sociology*, 65: 547–61.

Bell, D. and Valentine, G. (eds) (1995a) *Mapping Desire: Geographies of Sexualities*. London and New York: Routledge.

Bell, D. and Valentine, G. (1995b) 'Queer country: rural lesbian and gay lives', *Journal of Rural Studies*, 11: 113–22.

Bellah, R.N., Madsen, R., Sullivan, W.M., Swindler, A. and Tipton, S.M. (1985) *Habits of the Heart: Individualism and Commitment in American Life*. New York: Harper & Row.

Berg, L. (1994) 'Masculinity, place and a binary discourse of "theory" and "empirical investigation" in the human geography of Aotearoa/New Zealand', *Gender, Place and Culture*, 1: 245–60.

Berg, L.D. and Kearns, R.A. (1996) 'Naming as norming: "race", gender, and the identity politics of naming places in Aotearoa/New Zealand', *Environment and Planning D: Society and Space*, 14: 99–122.

Berg, L. and Kearns, R. (1998) 'Guest editorial: America unlimited', *Environment and Planning D: Society and Space*, 16: 128–32.

Berry, B.J.L. (1971) 'Monitoring trends, forecasting change and evaluating goals and achievements: the ghetto versus desegregation issues in Chicago as a case study', in Chisholm, M., Frey, W.E. and Haggett, P. (eds), *Regional Forecasting*. Colston Papers No. 22. London: Butterworth.

Berry, B.J.L. and Marble, D.F. (1968) 'Introduction', in Berry, B.J.L. and Marble, D.F. (eds), *Spatial analysis: A Reader in Statistical Geography*. Englewood Cliffs, NJ and London: Prentice-Hall. pp. 1–9.

Berry, M. (1981) 'Posing the housing question in Australia: elements of a theoretical framework for a Marxist analysis of housing', *Antipode*, 13: 2–14.

Bhaskar, R. (1978) *A Realist Theory of Science*. Brighton: Harvester.

Bhaskar, R. (1979) *The Possibility of Naturalism: A Philosophical Critique of the Contemporary Human Sciences*. Brighton: Harvester.

Bingham, N. and Thrift, N. (2000) 'Michael Serres and Bruno Latour', in Crang, M. and Thrift, N. (eds), *Thinking Space*. London: Routledge. pp. 281–301.

Binnie, J. (1997) 'Invisible Europeans: sexual citizenship in the new Europe', *Environment and Planning A*, 29: 237–48.

Binnie, J. and Valentine, G. (1999) 'Geographies of sexuality – a review of progress', *Progress in Human Geography*, 23: 175–87.

Blomley, N. (1994) 'Activism and the academy', *Environment and Planning D: Society and Space*, 12: 383–5.

Blumen, O. (2002) 'Criss-crossing boundaries: ultra orthodox Jewish women go to work', *Gender Place and Culture*, 9: 133–51.

Blunt, A. and Rose, G. (eds) (1994) *Writing Women and Space: Colonial and Postcolonial Geographies*. New York: Guilford Press.

Blunt, A. and Wills, J. (2000) *Dissident Geographies: An Introduction to Radical Ideas and Practice*. Harlow: Pearson Education.

Bondi, L. (1993) 'Locating identity politics', in Keith, M. and Pile, S. (eds), *Place and the Politics of Identity*. London: Routledge. pp. 84–101.

Bonnett, A. (1996) 'Constructions of "race", place and discipline: geographies of "racial" identity and racism', *Ethnic and Racial Studies*, 19: 864–83.

Bonnett, A. (1997) 'Geography, race and whiteness: invisible traditions and current challenges', *Area*, 29: 193–9.

Bookman, A. and Morgen, S. (1988) *Women and the Politics of Empowerment*. Philadelphia, PA: Temple University Press.

Bourdieu, P. (1984) *Distinction*. Cambridge, MA: Harvard University Press.

Bowlby, S.R., Foord, J., McDowell, L. and Momsen, J. (1982) 'Environment planning and feminist theory: a British perspective', *Environment and Planning A*, 14: 711–16.

Bowlby, S., Foord, J. and McDowell, L. (1986) 'The place of gender in locality studies', *Area*, 18: 327–31.

Bowlby, S., Lewis, J., McDowell, L. and Foord, J. (1989) 'The geography of gender', in Peet, R. and Thrift, N. (eds), *New Models in Geography*. Vol. 2. London: Unwin Hyman.

Brickell, C. (2000) 'Heroes and invaders: gay and lesbian pride parades and the public/private distinction in New Zealand media accounts', *Gender, Place and Culture*, 7: 163–78.

Bridge, G. (1995) 'The space for class? On class analysis in the study of gentrification', *Transactions of the Institute of British Geographers*, 20: 236–47.

Brown, M. (2000) *Closet Space: Geographies of Metaphor from the Body to the Globe*. London: Routledge.

Bruegel, I. (1973) 'Cities, women and social class: a comment', *Antipode*, 5: 62–3.

Bunge, W. (1977) 'The first years of the Detroit Geographical Expedition: a personal report', in Peet, R. (ed.), *Radical Geography*. London: Methuen. pp. 31–9.

Burnett, P. (1973) 'Social change, the status of women and models of city form and development', *Antipode*, 5: 57–61.

Butler, J. (1990) *Gender Trouble: Feminism and the Subversion of Identity*. New York: Routledge.

Butler, J. (1993) *Bodies that Matter: On the Discursive Limits of 'Sex'.* New York: Routledge.

Butler, J. (1997) *Excitable Speech*. London and New York: Routledge.

Butler, T. and Savage, M. (eds) (1995) *Social Change and the Middle Classes*. London: UCL Press.

Buttimer, A. (1976) 'Grasping the dynamism of the life-world', *Annals of the Association of American Geographers*, 66: 277–92.

Buttimer, A. (1979) 'Reason, rationality and human creativity', *Geografiska Annaler*, 61B: 43–9.

Byrne, D. (1998) 'Class and ethnicity in complex cities – the cases of Leicester and Bradford', *Environment and Planning A*, 30: 703–20.

Callon, M. (1986) 'Some elements of a sociology of translation', in Law, J. (ed.), *Power, Action and Belief: A New Sociology of Knowledge?* London: Routledge and Kegan Paul. pp. 234–65.

Cameron, J. (1998) 'The practice of politics: transforming subjectivities in the domestic domain and the public sphere', *Australian Geographer*, 29: 293–307.

The Canadian Geographer (1993) *The Canadian Geographer*, 37(1).

Capel, H. (1981) 'Institutionalisation of geography and strategies of change', in Stoddart, D.R. (ed.), *Geography, Ideology and Social Concern*. Oxford: Basil Blackwell. pp. 37–69.

Castells, M. (1983) *The City and the Grassroots*. Berkeley, CA: University of California Press.

Castree, N. (1999) 'Envisioning capitalism: geography and the renewal of Marxian political economy', *Transactions of the Institute of British Geographers*, 24: 137–58.

Castree, N. (2000) '"In here", "out there?" Domesticating critical geography', *Area*, 31: 81–6.

Childs, P. and Williams, W. (1997) *An Introduction to Post-colonial Theory*. London and New York: Prentice-Hall.

Chorley, R.J. and Haggett, P. (eds) (1967) *Models in Geography*. London: Methuen.

Chouinard, V. (1997) 'Structure and agency: contested concepts in human geography', *Canadian Geographer*, 41: 363–77.

Clark, G.L. (1989) *Unions and Communities under Siege: American Communities and the Crisis of Organized Labor*. New York: Cambridge University Press.

Clark, G.L. and Dear, M. (1984) *State Apparatus: Structures and Language of Legitimacy*. Boston, MA: Allen and Unwin.

Cloke, P. (1977) 'An index of rurality for England and Wales', *Regional Studies*, 11: 31–46.

Cloke, P. (1989) 'Rural geography and political economy', in Peet, R. and Thrift, N. (eds), *New Models in Geography*. London: Unwin and Hyman. pp. 164–97.

Cloke, P. (1994) '(En)culturing political economy: a life in the day of a "rural geographer"', in Cloke, P., Doel, M., Matless, D., Phillips, M. and Thrift, N. (eds), *Writing the Rural: Five Cultural Geographies*. London: Paul Chapman Publishing. pp. 149–90.

Cloke, P. (1995) 'Rural poverty and the welfare state: a discursive transformation in Britain and the USA', *Environment and Planning A*, 27: 1001–16.

Cloke, P. (1997) 'Country backwater to virtual village? Rural studies and "the cultural turn"', *Journal of Rural Studies*, 13: 367–75.

Cloke, P. (2002) 'Deliver us from evil? Prospects for living ethically and acting politically in human geography', *Progress in Human Geography*, 26: 587–604.

Cloke, P. and Edwards, G. (1986) 'Rurality in England and Wales 1981: a replication of the 1971 index', *Regional Studies*, 20: 289–306.

Cloke, P. and Goodwin, M. (1992) 'Conceptualising social change: from post-Fordism to rural structured coherence', *Transactions of the Institute of British Geographers*, NS, 17: 321–36.

Cloke, P. and Thrift, N. (1987) 'Intra-class conflict in rural areas', *Journal of Rural Studies*, 3: 321–33.

Cloke, P. and Thrift, N. (1990) 'Class and change in rural Britain', in Marsden, T., Lowe, P. and Whatmore, S. (eds), *Rural Restructuring*. London: David Fulton. pp. 165–81.

Cloke, P., Phillips, M. and Thrift, N. (1995) 'The new middle classes and the social constructs of rural living', in Butler, T. and Savage, M. (eds), *Social Change and the Middle Classes*. London: UCL Press. pp. 220–38.

Cloke, P., Philo, C. and Sadler, D. (1992) *Approaching Human Geography: An Introduction to Contemporary Theoretical Debates*. London: Paul Chapman.

Cloke, P., Milbourne, P. and Widdowfield, R. (2000) 'The hidden and emerging spaces of rural homelessness', *Environment and Planning A*, 32: 77–90.

Cohen, S. (1985) *Visions of Social Control: Crime, Punishment and Classification*. Cambridge: Polity Press.

Community Economies Collective (2001) 'Imagining and enacting noncapitalist futures', *Socialist Review*, 28: 93–135.

Connell, R.W. (1995) *Masculinities*. Berkeley, CA: University of California Press.

Connolly, W. (1995) *The Ethos of Pluralization*. Minneapolis, MN: University of Minnesota Press.

Cooke, P. (1985) 'Class practices as regional markers: a contribution to labour geography', in Gregory, D. and Urry, J. (eds), *Social Relations and Spatial Structures*. London: Macmillan. pp. 213–41.

Costello, L. and Hodge, S. (1999) 'Queer/clear/here: destabilising sexualities and spaces', in Stratford, E. (ed.), *Australian Cultural Geographies*. Melbourne: Oxford University Press. pp. 131–51.

Crang, M. (1998) *Cultural Geography*. London and New York: Routledge.

Cream, J. (1995) 'Re-solving riddles: the sexed body', in Bell, D. and Valentine, G. (eds), *Mapping Desire: Geographies of Sexualities*. London and New York: Routledge. pp. 31–40.

Crook, S., Pakulski, J. and Waters, M. (1992) *Postmodernization: Change in Advanced Society*. London: Sage.

Crush, J. (1991) 'The discourse of progressive human geography', *Progress in Human Geography*, 15: 395–414.

Crush, J. (1992) 'Power and surveillance on the South African gold mines', *Journal of Southern African Studies*, 18: 825–44.

Crush, J. (1994) 'Scripting the compound: power and space in the South African mining industry', *Environment and Planning D: Society and Space*, 12: 301–24.

Cybriwsky, R.A. (1978) 'Social aspects of neighbourhood change', *Annals of the Association of American Geographers*, 68: 17–33.

Dalby, S. and Mackenzie, F. (1997) 'Reconceptualizing local community: environment, identity and threat', *Area*, 29: 99–108.

Davies, B. (1991) 'The concept of agency: a feminist poststructuralist analysis', *Social Analysis*, 30: 42–53.

Davis, J. (1997) 'New Age Travellers in the countryside: incomers with attitude', in Milbourne, P. (ed.), *Revealing Rural 'Others': Representation, Power and Identity in the British Countryside*. London: Pinter. pp. 117–34.

Davis, S. and Prescott, J.R.V. (1992) *Aboriginal Frontiers and Boundaries in Australia*. Melbourne: Melbourne University Press.

Davis, T. (1995) 'The diversity of queer politics and the redefinition of sexual identity and community in urban spaces', in Bell, D. and Valentine, G. (eds), *Mapping Desire: Geographies of Sexualities*. London and New York: Routledge. pp. 284–303.

Day, K. (2001) 'Constructing masculinity and women's fear in public space in Irvine, California', *Gender Place and Culture*, 8: 109–27.

Dear, M. (1994) 'Postmodern human geography: a preliminary assessment', *Erdkunde*, 48: 2–13.

Delaney, D. and Leitner, H. (1997) 'The political construction of scale', *Political Geography*, 16: 93–7.

Dicken, P. and Thrift, N. (1992) 'The organization of production and the production of organisation: why business enterprises matter in the study of geographical industrialization', *Transactions of the Institute of British Geographers*, NS, 17: 279–91.

Driver, F. (1992) 'Geography's empire: histories of geographical knowledge', *Environment and Planning D: Society and Space*, 10: 23–40.

Driver, F. and Gilbert, D. (1998) 'Heart of empire? Landscape, space and performance in imperial London', *Environment and Planning D: Society and Space*, 16: 11–28.

Duce, R.H. (1935) (reprinted 1956) *Home and Overseas Geography (Regional Series) Book 5: A Simple World Survey – Africa*. London: Pitman.

Duncan, J.S. (1978) 'The social construction of unreality: an interactionist approach to the tourist's cognition of environment', in Ley, D. and Samuels, M. (eds), *Humanistic Geography: Prospects and Problems*. London: Croom Helm. pp. 269–82.

Duncan, N. (1996) *Bodyspace: Destabilizing Geographies of Gender and Sexuality*. London: Routledge.

Dunford, M. (1979) 'Capital accumulation and regional development in France', *Geoforum*, 10: 81–108.

Dunn, K.M., McGuirk, P.M. and Winchester, H.P.M. (1995) 'Place making: the social construction of Newcastle', *Australian Geographical Studies*, 33: 149–66.

Durie, E. (1987) 'The law and the land', in Phillips, J. (ed.), *Te Whenua, Te Iwi, the Land and the People*. Wellington: Allen and Unwin. pp. 78–81.

Dwyer, C. (1998) 'Contested identities: challenging dominant representations of young British Muslim women', in Skelton, T. and Valentine, G. (eds), *Cool Places: Geographies of Youth Cultures*. London: Routledge. pp. 50–65.

Dwyer, C. (1999) 'Veiled meanings: Young British Muslim women and the negotiation of difference', *Gender Place and Culture*, 6: 5–16.

Edel, M. (1982) 'Home ownership and working class unity', *International Journal of Urban and Regional Research*, 6: 205–22.

Edgell, S. (1993) *Class*. London: Routledge.

Elder, G. (1995) 'Of moffies, kaffirs and perverts: male homosexuality and the discourse of moral order in the apartheid state', in Bell, D. and Valentine, G. (eds), *Mapping Desire: Geographies of Sexualities*. London and New York: Routledge. pp. 56–65.

Elwood, S.A. (2000) 'Lesbian living spaces: multiple meanings of home', in Valentine, G. (ed.), *From Nowhere to Everywhere: Lesbian Geographies*. New York: Harrington Park Press. pp. 11–27.

England, K. (1991) 'Gender relations and the spatial structure of the city', *Geoforum*, 22: 135–47.

England, K.V.L. (1994) 'Getting personal: reflexivity, positionality and feminist research', *Professional Geographer*, 46: 80–9.

England, K. and Stiell, B. (1997) 'They think you are as stupid as your English is: constructing foreign domestic workers in Toronto', *Environment and Planning A*, 29: 195–215.

Entrikin, J.N (1976) 'Contemporary humanism in geography', *Annals of the Association of American Geographers*, 66: 615–32.

Entrikin, J.N. (1991) *The Betweenness of Place: Toward a Geography of Modernity*. Baltimore, MD: Johns Hopkins University Press.

Ettorre, E.M. (1978) 'Women, urban social movements and the lesbian ghetto', *International Journal of Urban and Regional Research*, 2: 449–520.

Ettorre, E.M. (1980) *Lesbians, Women and Society*. London: Routledge and Kegan Paul.

Fieldhouse, E.A. and Gould, M.I. (1998) 'Ethnic minority unemployment and local labour market conditions in Great Britain', *Environment and Planning A*, 30: 833–53.

Fincher, R. (1981) 'Analysis of the local level capitalist state', *Antipode*, 13: 25–3*.

Fincher, R. (1991) 'Caring for workers' dependents: gender, class and local practice in Melbourne', *Political Geography Quarterly*, 10: 356–81.

Fincher, R. and Jacobs, J. (eds) (1998) *Cities of Difference*. London: Guild

Fincher, R. and Panelli, R. (2001) 'Making space: women's urban and and the Australian state', *Gender Place and Culture*, 8: 129–48.

Fisher, C. (1997) '"I bought my first saw with my maternity ben tion in west Wales and the home as the space of (re)produc Little, J. (eds), *Contested Countryside Cultures: Otherne Rurality*. London: Routledge. pp. 232–51.

Foord, J. and Gregson, N. (1986) 'Patriarchy: towards a reconceptualisation', *Antipode*, 18: 186–211.

Ford, N., Halliday, J. and Little, J. (1999) 'Changes in the sexual lifestyles of young people in Somerset, 1990–1996', *British Journal of Family Planning*, 25: 55–8.

Forest, B. (1995) 'West Hollywood as symbol: the significance of place in the construction of a gay identity', *Environment and Planning D: Society and Space*, 13: 133–57.

Forrest, R. and Murie, A. (1987) 'The affluent homeowner: labour–market position and the shaping of housing histories', in Thrift, N. and Williams, P. (eds), *Class and Space: The Making of Urban Society*. London: Routledge and Kegan Paul. pp. 330–59.

Foucault, M. (1972) *The Archaeology of Knowledge*. Trans. A.M. Sheridan Smith. London: Tavistock.

Foucault, M. (1973) *The Birth of the Clinic: An Archaeology of Medical Perception*. London: Tavistock.

Foucault, M. (1979) *Discipline and Punish: The Birth of the Prison*. New York: Vintage Books.

Foucault, M. (1980) 'Two lectures', in Gordon, C. (ed.), *Power/Knowledge: Selective Interviews and Other Writings 1972–1977*. New York: Pantheon. pp. 78–108.

Foucault, M. (1981) 'The order of discourse', in Young, R. (ed.), *Untying the Text: A Poststructural Reader*. London: Routledge and Kegan Paul. pp. 48–79.

Foucault, M. (1990) *History of Sexuality: Volume 1*. London: Penguin.

Foucault, M. (1997) *Ethics: Subjectivity and Truth*. Ed. P. Rabinow, trans. R. Hurley. New York: The New Press.

Frobel, F., Heinrichs, J. and Kreye, O. (1980) *The New International Division of Labour*. Cambridge and New York: Cambridge University Press.

Gale, F. with Brookman, A. (1972) *Urban Aborigines*. Canberra: Australian University Press.

Gibson, C. (1998) '"We sing our home, we dance our land": indigenous self-determination and contemporary geopolitics in Australian popular music', *Environment and Planning D: Society and Space*, 16: 163–84.

Gibson, K. (1992) '"Hewers of cake and drawers of tea": women, industrial restructuring and class processes on the coalfields of central Queensland', *Rethinking Marxism*, 5: 29–56.

Gibson, K. (1993) *Different Merry-go-rounds: Families, Communities and the 7-day Roster*. Brisbane: Miners Federation, Queensland Branch.

Gibson, K. (1998) 'Social polarization and the politics of difference: discourses in collision or collusion?', in Fincher, R. and Jacobs, J.M. (eds), *Cities of Difference*. New York: Guilford Press. pp. 301–16.

Gibson, K. and Graham, J. (1992) 'Rethinking class in industrial geography', *Economic Geography*, 68: 109–27.

Gibson, K. and Watson, S. (eds) (1996) *Postmodern Cities and Spaces*. Oxford: Basil Blackwell.

Gibson, K., Graham, J. and Shakow, D. (1989) 'Calculating economic indicators in value terms: the Australian economy and industrial sectors, 1974–75 and 1978–79', *Journal of Australian Political Economy*, 25: 17–43.

Gibson-Graham, J.K. (1994) '"Stuffed if I know!" Reflections on postmodern feminist social research', *Gender, Place and Culture*, 1: 205–24.

Gibson-Graham, J.K. (1995) 'Beyond patriarchy and capitalism: reflections of political subjectivity', in Caine, B. and Pringle, R. (eds), *Transitions: New Australian Feminisms*. Sydney: Allen and Unwin. pp. 172–83.

Gibson-Graham, J.K. (1996a) *The End of Capitalism (as we knew it): A Feminist Critique of Political Economy*. Cambridge, MA: Blackwell.

Gibson-Graham, J.K. (1996b) 'Reflections on post-modern feminist social research', in Duncan, N. (ed.), *Bodyspace: Destabilizing Geographies of Gender and Sexuality*. London: Routledge. pp. 234–44.

Gibson-Graham, J.K. (2000) 'Poststructural interventions', in Sheppard, E. and Barnes, T. (eds), *A Companion to Economic Geography*. Oxford and Malden, MA: Blackwell. pp. 95–110.

Gibson-Graham, J.K. (2003) 'An ethics of the local', *Rethinking Marxism*, 15.

Giddens, A. (1973) *The Class Structure of the Advanced Societies*. London: Hutchinson.

Giddens, A. (1984) *The Constitution of Society: Outline of the Theory of Structuration*. Cambridge: Polity Press.

Gilbert, M.R. (1998) '"Race", space and power: the survival strategies of working poor women', *Annals of the Association of American Geographers*, 88: 595–621.

Gleeson, B. (1999) *Geographies of Disability*. London: Routledge.

Gleeson, B. and Kearns, R. (2001) 'Remoralising landscapes of care', *Environment and Planning D: Society and Space*, 19: 61–80.

Goldthorpe, J.H. (1982) 'On the service class, its formation and future', in Giddens, A. and MacKenzie, G. (eds), *Social Class and the Division of Labour*. Cambridge: Cambridge University Press.

Gonzalez, J. and Habel-Pallan, M. (1994) 'Heterotopias and shared methods of resistance: navigating social spaces and spaces of identity', *Inscriptions*, 7: 80–104.

Gough, J. (2001) 'Work, class and social life', in Pain, R. with Barke, M., Fuller, D., Gough, J., MacFarlane, R. and Mowl, G. (eds), *Introducing Social Geographies*. London: Arnold. pp. 13–43.

Graham, J. (1990) 'Theory and essentialism in Marxist geography', *Antipode*, 22: 53–66.

Graham, J. (1992) 'Anti-essentialism and over determination – a response to Dick Peet', *Antipode*, 24: 141–56.

Graham, J., Gibson, K., Horvath, R. and Shakow, D. (1988) 'Restructuring in U.S. manufacturing: the decline of monopoly capitalism', *Annals of the Association of American Geographers*, 78: 473–90.

Grant, A. (2000) 'And still, the lesbian threat: or, how to keep a good woman a woman', in Valentine, G. (ed.), *From Nowhere to Everywhere: Lesbian Geographies*. New York: Harrington Park Press. pp. 61–80.

Gregory, D. (1989) Presences and absences: time–space relations and structuration theory', in Held, D. and Thompson, J.B. (eds), *Social Theory of the Modern Societies: Anthony Giddens and His Critics*. Cambridge: Cambridge University Press. pp. 185–214.

Gregory, D. and Urry, J. (eds) (1985) *Social Relations and Spatial Structures*. London: Macmillan.

Gregson, N. (1989) 'On the (ir)relevance of structuration theory to empirical research', in Held, D. and Thompson, J.B. (eds), *Social Theory of the Modern*

Societies: Anthony Giddens and His Critics. Cambridge: Cambridge University Press. pp. 235–48.

Gregson, N. and Rose, G. (2000) 'Taking Butler elsewhere: performances, spatialities and subjectivities', *Environment and Planning D: Society and Space*, 18: 433–52.

Guelke, L. (1978) 'Geography and logical positivism', in Herbert, D.T. and Johnston, R.J. (eds), *Geography and the Urban Environment*. Vol. 1. New York: John Wiley. pp. 35–61.

Habermas, J. (1978) *Knowledge and Human Interests*. London: Heinemann.

Haggett, P. (1965) *Locational Analysis in Human Geography*. London: Edward Arnold.

Halfacree, K. (1993) 'Locality and social representation: space, discourse and alternative definitions of the rural', *Journal of Rural Studies*, 9: 1–15.

Hall, S. (1996) 'Introduction: Who needs 'identity'?', in Hall, S. and du Gay, P. (eds), *Questions of Cultural Identity*. London: Sage. pp. 1–17.

Halliday, J. and Little, J. (2001) 'Amongst women: exploring the reality of rural childcare', *Sociologia Ruralis*, 41: 423–37.

Halsey, A.H., Heath, A.F. and Ridge, J.M. (1980) *Origins and Destinations: Family, Class and Education in Modern Britain*. Oxford: Clarendon Press.

Hamnett, C. (1996) *Social Geography: A Reader*. London: Arnold.

Hamnett, C. (2003) 'Contemporary human geography: fiddling while Rome burns?', *Geoforum*, 34: 1–3.

Hansen, P. (1999) 'Guides and sherpas in the Alps and Himalayas', in Elsner, J. and Rubies, J. (eds), *Voyages and Visions: Toward a Cultural History of Travel*. London: Reaktion Books. pp. 210–31.

Hanson, S. (1992) 'Geography and feminism: worlds in collision?', *Annals of the Association of American Geographers*, 82: 569–86.

Hanson, S. (2000) 'Focus section: women in geography in the 21st century networking', *The Professional Geographer*, 52: 751–8.

Hanson, S. and Pratt, G. (1988) 'Spatial dimensions of the gender division of labor in a local labor market', *Urban Geography*, 9: 180–202.

Hanson, S. and Pratt, G. (1995) *Gender, Work and Space*. London and New York: Routledge.

Harrington, V. and O'Donoghue, D. (1998) 'Rurality in England and Wales 1991: a replication and extension of the 1981 Rurality Index', *Sociologia Ruralis*, 38(2): 178–203.

Harrison, P. (2000) 'Making sense: embodiment and the sensibilities of the everyday', *Environment and Planning D: Society and Space*, 18: 497–517.

Hartsock, N. (1987) 'Rethinking modernism: minority versus majority theories', *Cultural Critique*, 7: 187–206.

Harvey, D. (1969) *Explanation in Geography*. London: Edward Arnold.

Harvey, D. (1973) *Social Justice and the City*. London: Edward Arnold.

Harvey, D. (1976) 'The Marxist theory of the state', *Antipode*, 8: 80–9.

Harvey, D. (1977) 'The geography of capitalist accumulation: a reconstruction of the Marxian theory', in Peet, R. (ed.), *Radical Geography*. Chicago: Maaroufa. pp. 263–92.

Harvey, D. (1982) *The Limits to Capital*. Oxford and Cambridge, MA: Blackwell.

Harvey, D. (1984) 'On the history and present conditions of geography: an historical materialist manifesto', *Professional Geographer*, 3: 1–11.

Harvey, D. (1989) 'From models to Marx: notes on the project to remodel contemporary geography', in McMillan, B. (ed.), *Remodelling Geography*. Oxford: Basil Blackwell. pp. 211–16.

Hay, I. (1995) 'The strange case of Dr Jeckyll on Hyde Park: fear, media and the conduct of emancipatory geography', *Australian Geographical Studies*, 33: 257–71.

Hay, I. (1998) 'Making moral imaginations: research ethics, pedagogy, and professional human geography', *Ethics, Place and Environment*, 1: 55–75.

Hayford, A. (1974) 'The geography of women: an historical introduction, *Antipode*, 6: 1–19.

Hemmings, C. (1995) 'Locating bisexual identities: discourses of bisexuality and contemporary feminist theory', in Bell, D. and Valentine, G. (eds), *Mapping Desire: Geographies of Sexualities*. London and New York: Routledge. pp. 41–55.

Herbert, D.T. (1968) 'Principal component analysis and British studies of urban–social structure', *The Professional Geographer*, 20: 280–3.

Herod, A. (1991a) 'Local political practice in response to a manufacturing plant closure: how Geography complicates class analysis', *Antipode*, 23: 385–402.

Herod, A. (1991b) 'The production of scale in United States labour relations', *Area*, 23: 82–8.

Herod, A. (1992) *Towards a Labor Geography: The Production of Space and the Politics of Scale in the East Coast Longshore Industry, 1953–1990*. Unpublished PhD thesis, Department of Geography, Rutgers University, New Brunswick, NJ.

Herod, A. (1993) 'Gender issues in the use of interviewing as a research method', *The Professional Geographer*, 45: 305–17.

Herod, A. (1994) 'Further reflections on organized labor and deindustrialization in the United States', *Antipode*, 26: 77–95.

Herod, A. (1995) 'The practice of international labor solidarity and the geography of the global economy', *Economic Geography*, 71: 341–63.

Herod, A. (1998a) 'Discourse on the docks: containerization and inter-union work disputes in US ports, 1955–1985', *Transactions of the Institute of British Geographers*, NS, 23: 177–91.

Herod, A. (1998b) 'Of blocs, flows and networks: the end of the Cold War, cyberspace, and the geo-economics of organized labor at the *fin de millénaire*', in Herod, A., Ó Tuathail, G. and Roberts, S. (eds), *An Unruly World? Globalization, Governance and Geography*. London: Routledge. pp. 162–95.

Herod, A. (1998c) 'The geostrategics of labor in post-Cold War Eastern Europe: an examination of the activities of the International Metalworkers' Federation', in Herod, A. (ed.), *Organizing the Landscape: Geographical Perspectives on Labor Unionism*. Minneapolis, MN: University of Minnesota Press. pp. 45–74.

Herod, A. (2000) 'Workers and workplaces in a neoliberal global economy', *Environment and Planning A*, 32: 1781–90.

Herod, A. (2001a) *Labor Geographies: Workers and the Landscapes of Capitalism*. New York: Guilford Press.

Herod, A. (2001b) 'Labor internationalism and the contradictions of globalization: or, why the local is sometimes still important in a global economy', *Antipode*, 33: 407–26.

Hetherington, K. (1998) *Expressions of Identity: Space, Performance, Politics*. London: Sage.

Hetherington, K. (2000) *New Age Travellers: Vanloads of Uproarious Humanity*. London and New York: Cassell.

Hetherington, K. and Law, J. (eds) (2000) 'Special issue: after networks', *Environment and Planning D: Society and Space*, 18, issue 2.

Holloway, J. and Valins, O. (2002) 'Editorial and special issue: placing religion and spirituality in geography', *Social and Cultural Geography*, 3: 5–9.

Holloway, S.R. (1998) 'Male youth activities and metropolitan context', *Environment and Planning A*, 30: 385–99.

Holloway, S.R. and Valentine, G. (2000) *Children's Geographies: Playing, Living, Learning*. London: Routledge.

Holt-Jensen, A. (1999) *Geography: History and Concepts, A Students Guide*, 3rd edn. London: Sage.

hooks, b. (1990) *Yearnings: Race, Gender, Cultural Politic*. Boston, MA: South End Press.

hooks, b. (1992) *Teaching to Transgress: Education as the Practice of Freedom*. London: Routledge.

Hubbard, P.J. (1998) 'Community action and the displacement of street prostitutes: evidence from British cities', *Geoforum*, 29: 269–86.

Hubbard, P.J. (2000) 'Desire/disgust: moral geographies of heterosexuality', *Progress in Human Geography*, 24: 191–217.

Hubbard, P.J. (2001) 'Sex zones: desire, citizenship and public space', *Sexualities*, 4: 51–71.

Hubbard, P. (2002) 'Sexing the Self: geographies of engagement and encounter', *Journal of Social and Cultural Geography*, 3: 365–81.

Hubbard, P., Kitchin, R., Bartley, B. and Fuller, D. (2002) *Thinking Geographically: Space, Theory and Contemporary Human Geography*. London: Continuum.

Hughes, A. (1997) 'Rurality and cultures of womanhood: domestic identities and moral order in village life', in Cloke, P. and Little, J. (eds), *Contested Countryside Cultures: Otherness, Marginalisation and Rurality*. London: Routledge. pp. 123–37.

Jackson, P. (1984) 'Social disorganisation and moral order in the city', *Transactions of the Institute of British Geographers*, NS, 9: 168–80.

Jackson, P. (1991) 'The cultural politics of masculinity: towards a social geography', *Transaction of the Institute of British Geographers*, NS, 16: 199–213.

Jackson, P. (1992) *Maps of Meaning: An Introduction to Cultural Geography*. London and New York: Routledge.

Jackson, P. (1994) 'Black male: advertising and the cultural politics of masculinity', *Gender, Place and Culture*, 1: 49–59.

Jackson, P. (1998) 'Constructions of "whiteness" in the geographical imagination', *Area*, 30: 99–106.

Jackson, P. (1999) 'Identity', in McDowell, L. and Sharp, J. (eds), *A Feminist Glossary of Human Geography*. London: Arnold. pp. 132–4.

Jackson, P. (2000) 'Rematerializing social and cultural geography', *Social and Cultural Geography*, 1: 9–14.

Jackson, P. and Jacobs, J. (1996) 'Postcolonialism and the politics of race', *Environmental and Planning D: Society and Space*, 14: 1–3.

Jackson, P. and Penrose, J. (1994) *Constructions of Race, Place and Nation*. Minneapolis, MN: Minnesota University Press.

Jackson, P. and Smith, S. (1984) *Exploring Social Geography*. London: Allen and Unwin.

Jackson, P., Stevenson, N. and Brooks, K. (1999) 'Making sense of men's lifestyle magazines', *Environment and Planning D: Society and Space*, 17: 353-68.

Jacobs, J. (1994) '"Shake 'im this country": the mapping of the Aboriginal sacred in Australia - the case of Coronation Hill', in Jackson, P. and Penrose, J. (eds), *Constructions of Race, Place and Nation*. Minneapolis, MN: Minnesota University Press. pp. 100-18.

Jacobs, J. (1996) *Edge of Empire: Postcolonialism and the City*. London and New York: Routledge.

Jagose, A. (1996) *Queer Theory*. Melbourne: Melbourne University Press.

Jarosz, L. and Lawson, V. (2002) '"Sophisticated people versus rednecks": economic restructuring and class difference in America's West', *Antipode*, 34: 8-27.

Johnson, L. (1987) '(Un) Realist perspectives: patriarchy and feminist challenges in geography', *Antipode*, 19: 210-15.

Johnson, L. (1991) The Australian Textile Industry, 1965-1990: A Feminist Geography. Unpublished PhD thesis, Department of Geography, Monash University, Melbourne.

Johnson, L., with Huggins, J. and Jacobs, J. (2000) *Placebound: Australian Feminist Geographies*. Melbourne: Oxford University Press.

Johnston, L. (1995) 'The politics of the pump: hard-core gyms and women body builders', *New Zealand Geographer*, 51: 16-18.

Johnston, L. (1996) 'Pumped-up politics: female body builders refiguring the body', *Gender, Place and Culture: A Journal of Feminist Geography*, 3(3): 327-40.

Johnston, L. (1997) 'Queen(s') Street or Ponsonby poofters? The embodied HERO parade site', *New Zealand Geographer*, 53: 29-33.

Johnston, L. (1998) 'Reading sexed bodies in sexed spaces', in Nast, H. and Pile, S. (eds), *Places Through the Body*. London: Routledge.

Johnston, L. (2002) 'Borderline bodies at gay pride parades', in Feminist Geography Reading Group, *Subjectivities, Knowledges and Feminist Geographies*. London: Rowman and Littlefield.

Johnston, L. and Valentine, G. (1995) 'Where ever I lay my girlfriend that's my home: performance and surveillance of lesbian identity in home environments', in Bell, D. and Valentine, G. (eds), *Mapping Desires: Geographies of Sexualities*. London: Routledge. pp. 149-66.

Johnston, R.J. (1991) *Geography and Geographers: Anglo-American Human Geography since 1945*. London: Edward Arnold.

Johnston, R.J. (1996) 'Academic tribes, disciplinary containers, and the real politic of opening up the social sciences', *Environment and Planning A*, 28: 1943-7.

Johnston, R.J. (2000) 'Authors, editors and authority in the postmodern academy', *Antipode*, 32: 271-91.

Johnston, R.J., Gregory, D., Pratt, G. and Watts, M. (eds) (2000) *The Dictionary of Human Geography*. Oxford: Blackwell.

Jones, D. and Womack, J. (1985) 'Developing countries and the future of the world automobile industry', *World Development*, 13: 393-408.

Jones, J.P., Nast, H. and Roberts, S. (eds) (1997) *Thresholds in Feminist Geography: Difference, Methodology, Representation*. Lanham, MD: Rowman and Littlefield.

Jones, O. (1999) 'Tomboy tales: the rural, nature and the gender of childhood', *Gender, Place and Culture*, 6: 117-36.

Kearns, R. (1997) 'Constructing (bi)cultural geographies: research on, and with, people of the Hokianga district', *New Zealand Geographer*, 53: 3–8.

Kearns, R.A. and Berg, L.D. (2002) 'Proclaiming place: towards a geography of place name pronunciation', *Social and Cultural Geography*, 3: 283–302.

Keith, M. (1997) 'Conclusion: a changing space and a time for change', in Pile, S. and Keith, M. (eds), *Geographies of Resistance*. London: Routledge. pp. 277–86.

Keith, M. and Pile, S. (eds) (1993) *Place and the Politics of Identity*. London: Routledge.

Kitchin, R. (1998) *Cyberspace: The World in the Wires*. Chichester: John Wiley.

Kitchin, R. and Tate, N.J. (2000) *Conducting Research into Human Geography: Theory, Methodology and Practice*. Harlow: Prentice Hall.

Knopp, L. (1987) 'Social theory, social movements and public policy: recent accomplishments of the gay and lesbian movements in Minneapolis, Minnesota', *International Journal of Urban and Regional Research*, 11: 243–61.

Knopp, L. (1990a) 'Exploiting the rent-gap: the theoretical significance of using illegal appraisal schemes to encourage gentrification in New Orleans', *Urban Geography*, 11: 48–64.

Knopp, L. (1990b) 'Some theoretical implications of gay involvement in an urban land market', *Political Geography Quarterly*, 9: 337–52.

Knopp, L. (1992) 'Sexuality and the spatial dynamics of capitalism', *Environment and Planning D: Society and Space*, 10: 671–89.

Knopp, L. (1994) 'Social justice, sexuality and the city', *Urban Geography*, 15: 644–60.

Knopp, L. (1995) 'Sexuality and urban space: a framework for analysis', in Bell, D. and Valentine, G. (eds), *Mapping Desire: Geographies of Sexualities*. London and New York: Routledge. pp. 149–61.

Knopp, L. (1997) 'Rings, circles and perverted justice: gay judges and moral panic in contemporary Scotland', in Keith, M. and Pile, S. (eds), *Geographies of Resistance*. London and New York: Routledge. pp. 168–83.

Knopp, L. (1998) 'Sexuality and urban space: gay male identity politics in the United States, the United Kingdom and Australia', in Fincher, R. and Jacobs, J.M. (eds), *Cities of Difference*. New York: Guilford Press. pp. 149–76.

Knopp, L. (1999) 'Out in academia: the queer politics of one geographer's sexualisation', *Journal of Geography in Higher Education*, 23: 116–23.

Knopp, L. (2000), 'A queer journey to queer geography', in Moss, P. (ed.), *Engaging Autobiography: Geographers Writing Lives*. Syracuse, NY: Syracuse University Press. pp. 78–98.

Knopp, L. and Brown, M.P. (2002) 'We're here! We're queer! We're over there too! Queer cultural geographies', in Anderson, K., Domosh, M., Pile, S. and Thrift, N. (eds), *The Handbook of Cultural Geography*. New York: Sage. pp. 460–81.

Knopp, L. and Brown, M.P. (forthcoming) 'Queer diffusions', *Environment and Planning D: Society and Space*.

Knopp, L. and Lauria, M. (1987) 'Gender relations as a particular form of social relations', *Antipode*, 19: 48–53.

Kramer, J.L. (1995) 'Bachelor farmers and spinsters: gay and lesbian identities and communities in rural North Dakota', in Bell, D. and Valentine, G. (eds), *Mapping Desire: Geographies of Sexualities*. London and New York: Routledge. pp. 200–13.

Laclau, E. and Mouffe, C. (1985) *Hegemony and Socialist Strategy: Towards a Radical Democratic Politics*. Trans. W. Moore and P. Cammack. London: Verso.

Latour, B. (1991) 'Technology is society made durable', in Law, J. (ed.), *A Sociology of Monsters. Essays on Power, Technology and Domination*. London: Routledge. pp. 103–30.

Lauria, M. and Knopp, L. (1985) 'Toward an analysis of the role of gay communities in the urban renaissance', *Urban Geography*, 6: 152–69.

Laurie, N., Dwyer, C., Holloway, S. and Smith, F. (eds) (1999) *Geographies of New Femininities*. London: Longman.

Law, J. (1994) *Organizing Modernity*. Oxford: Blackwell.

Law, R. (1997) 'Masculinity, place and beer advertising in New Zealand: the Southern Man campaign', *New Zealand Geographer*, 53: 22–8.

Leckie, G.J. (1993) 'Female farmers in Canada and the gender relations of a restructuring agricultural system', *The Canadian Geographer*, 37: 212–30.

Lewins, F. (1992) *Social Science Methodology: A Brief but Critical Introduction*. Melbourne: Macmillan.

Ley, D. (1974) *The Black Inner City and Frontier Outpost: Images and Behaviour of a Philadelphia Neighbourhood*. Washington, DC: Association of American Geographers.

Ley, D. (1978) 'Social geography and social action', in Ley, D. and Samuels, S. (eds), *Humanistic Geography: Prospects and Problems*. Chicago: Maaroufa Press. pp. 41–57.

Ley, D. (1980) 'Liberal ideology and the post-industrial city', *Annals of the Association of American Geographers*, 70: 238–58.

Ley, D. (1981) 'Behavioural geography and the philosophy of meanings', in Cox, K.R. and Golledge, R.G. (eds), *Behavioural Problems in Geography Revisited*. London: Methuen. pp. 209–30.

Ley, D. (1987) 'Styles of the times: liberal and neo-conservative landscapes in inner Vancouver, 1968–1986', *Journal of Historical Geography*, 13: 40–56.

Ley, D. (1994) 'Gentrification and the politics of the new middle class', *Environment and Planning D: Society and Space*, 12: 53–74.

Ley, D. (1996) *The New Middle Class and the Remaking of the Inner City*. Oxford: Oxford University Press.

Ley, D. and Samuels, S. (eds) (1978) *Humanistic Geography: Prospects and Problems*. Chicago: Maaroufa Press.

Liepins, R. (1996) 'Reading agricultural power: media as sites and processes in the production of meaning', *New Zealand Geographer*, 52: 3–10.

Liepins, R. (1998a) 'Fields of action: Australian women's agricultural activism in the 1990s', *Rural Sociology*, 63: 128–56.

Liepins, R. (1998b) 'The gendering of farming and agricultural politics: a matter of discourse and power', *Australian Geographer*, 29: 371–88.

Liepins, R. (1998c) '"Women of broad vision": nature and gender in the environmental activism of Australia's 'Women in Agriculture' movement', *Environment and Planning A*, 30: 1179–96.

Liepins, R. (1999) 'Women in agriculture: action for more democratic Australian farm politics', in Bystydzienski, J. and Sekhon, J. (eds), *Democratization and Women's Grassroots Movements*. Bloomington, IN: Indiana University Press. pp. 352–74.

Liepins, R. (2000a) 'Exploring rurality through "community": discourses, practices and spaces shaping Australian and New Zealand rural "communities"', *Journal of Rural Studies*, 16: 83–99.

Liepins, R. (2000b) 'Making men: the construction and representation of agriculture-based masculinities in Australia and New Zealand', *Rural Sociology*, 65: 605–20.

Liepins, R. (2000c) 'New energies for an old idea: reworking approaches to "community" in contemporary rural studies', *Journal of Rural Studies*, 16: 23–35.

Liepins, R. and Bradshaw, B. (1999) 'Neoliberal agricultural discourse in New Zealand: economy, culture and politics linked', *Sociologia Ruralis*, 39: 563–82.

Little, J. (1986) 'Feminist perspectives in rural geography: an introduction', *Journal of Rural Studies*, 2: 1–8.

Little, J. (1987) 'Gender relations in rural areas: the importance of women's domestic role', *Journal of Rural Studies*, 3: 335–42.

Little, J. (1991a) 'A rural idyll? Women and planning in rural areas', *Women and the Built Environment*, 15 (no numbers).

Little, J. (1991b) 'Theoretical issues of women's non-agricultural employment in rural areas', *Journal of Rural Studies*, 7: 99–105.

Little, J. (1994a) *Gender, Planning and the Policy Process*. Oxford: Butterworth-Heinneman.

Little, J. (1994b) 'Gender relations and the rural labour process', in Whatmore, S., Marsen, T. and Lowe, P. (eds), *Gender and Rurality*. London: David Fulton. pp. 11–30.

Little, J. (1994c) 'Women's initiatives in local authority planning departments in England', *Town Planning Review*, 65: 261–76.

Little, J. (1997) 'Employment, marginality and women's self-identity', in Cloke, P. and Little, J. (eds), *Contested Countryside Cultures: Otherness, Marginalisation and Rurality*. London: Routledge. pp. 138–57.

Little, J. (1999) 'Gender and rural policy', in Greed, C. (ed.), *Social Town Planning*. London: Routledge.

Little, J. (2001) *Gender and Rural Geography: Identity, Sexuality and Power in the Countryside*. London: Prentice-Hall.

Little, J. (2002a) 'Riding the love train: rural community and the heterosexual family'. Paper presented at the annual conference of the Rural Economy and Society Study Group, University of Cardiff: Cardiff.18–19 September.

Little, J. (2002b) 'Rural geography: rural gender identity and the performance of masculinity and femininity in the countryside', *Progress in Human Geography*, 26: 665–70.

Little, J. (forthcoming) 'Riding the love train: heterosexuality and the rural community', *Sociologia Ruralis*.

Little, J. and Austin, P. (1996) 'Women and the rural idyll', *Journal of Rural Studies*, 12: 101–11.

Little, J. and Jones, O. (2000) 'Masculinities, gender and rural policy', *Rural Sociology*, 65: 621–39.

Little, J. and Leyshon, M. (2003) 'Geographies of embodiment: a rural perspective', *Progress in Human Geography*, 27: 257–72.

Little, J., Peake, L. and Richardson, P. (eds) (1988) *Women in Cities: Gender and the Urban Environment*. Basingstoke: Macmillan.

Livingstone, D.N. (1992) *The Geographical Tradition: Episodes in a Contested Enterprise*. Oxford: Blackwell.

Lo, J. and Healy, T. (2000) 'Flagrantly flaunting it? Contesting perceptions of locational identity among urban Vancouver lesbians', in Valentine, G. (ed.), *From Nowhere to Everywhere: Lesbian Geographies*. New York: Harrington Park Press. pp. 29–44.

Longhurst, R. (1995) 'Geography and the body', *Gender, Place and Culture*, 2: 97–105.

Longhurst, R. (2000a) '"Corporeographies" of pregnancy: "bikini babes"', *Environment and Planning D: Society and Space*, 18: 453–72.

Longhurst, R. (2000b) 'Geography and gender: masculinities, male identity and men', *Progress in Human Geography*, 24: 439–44.

Longhurst, R. (2001) *Bodies: Exploring Fluid Boundaries*. London: Routledge.

Longhurst, R. (2002) 'Geography and gender: a "critical" time?', *Progress in Human Geography*, 26: 544–52.

Longhurst, R. and Johnston, L. (1998) 'Embodying places and emplacing bodies: pregnant women and women body builders', in De Plessis, R. and Alice, L. (eds), *Feminist Thought in Aotearoa/New Zealand: Connections and Differences*. Auckland: Oxford University Press.

Loyd, B. and Rowntree, L. (1978) 'Radical feminists and gay men in San Francisco: social place in dispersed communities', in Lanegrand, D. and Palm, R. (eds), *Invitation to Geography*. New York: McGraw-Hill. pp. 78–88.

Luzzadder-Beach, S. and Macfarlane, A. (2000) 'The environment of gender and science: status and perspectives of women and men in physical geography', *The Professional Geographer*, 52: 407–24.

Lyon, D. (1991) 'Bentham's panopitcan: from moral architecture to electronic surveillance', *Queens Quarterly*, 98: 596–617.

McAdam, D., McCarthy, J.D. and Zald, M.N. (eds) (1996) *Comparative Perspectives on Social Movements: Political Opportunities, Mobilizing Structures, and Cultural Framings*. Cambridge: Cambridge University Press.

McCarthy, J.D. and Zald, M.N. (1973) *The Trend of Social Movements in America: Professionalization and Resource Mobilization*. Morristown, NJ: General Learning Press.

McCarthy, J.D. and Zald, M.N. (1977) 'Resource mobilization and social movements: a partial theory', *American Journal of Sociology*, 82: 1212–41.

McClean, R., Berg, L.D. and Roche, M.M. (1997) 'Responsible geographies: co-creating knowledge in Aotearoa', *New Zealand Geographer*, 53: 9–15.

MacDonald, M. (2002) 'Towards a spatial theory of worship: some observations from Presbyterian Scotland', *Social and Cultural Geography*, 3: 61–80.

McDowell, L. (1979) 'Women in British Geography', *Area*, 11: 151–5.

McDowell, L. (1983) 'Towards an understanding of the gender divisions of urban space', *Environment and Planning D: Society and Space*, 1: 59–72.

McDowell, L. (1986) 'Beyond patriarchy: a class-based explanation of women's subordination', *Antipode*, 18: 311–21.

McDowell, L. (1990) 'Sex and power in academia', *Area*, 22: 323–32.

McDowell, L. (1992) 'Doing gender: feminism, feminists and research methods in human geography', *Transactions of the Institute of British Geographers*, NS, 17: 399–416.

McDowell, L. (1995) 'Body work: heterosexual gender performances in city workplaces', in Bell, D. and Valentine, G. (eds), *Mapping Desire: Geographies of Sexualities.* London and New York: Routledge.

McDowell, L. (1999) *Gender, Identity and Place: Understanding Feminist Geographies.* Cambridge: Polity Press.

McDowell, L. (2000) 'Father and Ford revisited: gender, class and employment change in the new millennium', *Transactions of the Institute of British Geographers,* 26: 448-64.

McDowell, L. (2002a) 'Masculine discourses and dissonances: strutting "lads", protest masculinity and domestic respectability', *Environment and Planning D: Society and Space,* 20: 97-119.

McDowell, L. (2002b) 'Transitions to work: masculine identities, youth inequality and labour mark change', *Gender Place and Culture,* 9: 39-59.

McDowell, L. and Massey, D. (1984) 'A women's place', in Massey, D. and Allen, J. (eds), *Geography Matters! A Reader.* Cambridge: Cambridge University Press. pp. 128-47.

McGuirk, P. (2000) 'Power and policy networks in urban governance: local government and property-led regeneration in Dublin', *Urban Studies,* 37: 651-72.

Mackenzie, F. (1992a) 'The politics of partnership: farm women and farm land, Ontario', in Bowler, I.R., Bryant, C.R. and Nellis, M.D. (eds), *Contemporary Rural Studies in Transition: Economy and Society.* Vol. 2. Wallingford: CAB International. pp. 85-95.

Mackenzie, F. (1992b) 'The worse it got the more we laughed: a discourse of resistance among farmers of Eastern Ontario', *Environment and Planning D: Society and Space,* 10: 691-713.

Mackenzie, F. (1994) 'Is where I sit, where I stand? The Ontario farm women's network, politics and difference', *Journal of Rural Studies,* 10: 101-15.

Mackenzie, S. (1984) 'Editorial', *Antipode,* 16: 3-9.

Mackenzie S. (1988) 'Women in the city', in Peet, R. and Thrift, N. (eds), *New Models in Geography.* Vol. 2. London: Unwin Hyman. pp. 109-26.

Mackenzie, S. and Rose, D. (1983) 'Industrial changes, the domestic economy and home life', in Anderson, J., Duncan, S. and Hudson, R. (eds), *Redundant Spaces: Industrial Decline in Cities and Regions.* London: Macmillan.

Mackenzie, S., Foord, J. and McDowell, L. (1980) 'Women's place, women's space: comments arising from papers in *Area'*, *Area,* 12: 47-51.

McLafferty, S.L. (2002) 'Mapping women's worlds: knowledge, power and the bounds of GIS', *Gender Place and Culture,* 9: 263-9.

MacLaughlin, J. (1998) 'The political geography of anti-Traveller racism in Ireland: the politics of exclusion and the geography of closure', *Political Geography,* 17: 417-35.

Maddrell, A.M.C. (1998) 'Discourses of race and gender and the comparative method in geography school texts 1830-1918', *Environment and Planning D: Society and Space,* 16: 81-103.

Mann, M. (1986) *The Sources of Social Power, Vol. 1: A History of Power from the Beginning to AD 1760.* Cambridge: Cambridge University Press.

Marsden, T., Murdoch, J., Lowe, P., Munton, R. and Flynn, A. (1993) *Constructing the Countryside.* London: UCL Press.

Marshall, J.U. (1985) 'Geography as a scientific enterprise', in Johnston, R.J. (ed.), *The Future of Geography.* London: Methuen. pp. 113-28.

Martin, R. (2001). 'Editorial: the geographer as social critic – getting indignant about income inequality', *Transactions of the Institute of British Geographers*, 26: 267–72.

Marx, K. (1967) *Capital. Vol. 3: The Process of Capitalist Production as a Whole.* New York: International Publishers.

Massey, D. (1983) 'Industrial restructuring as class restructuring: production, decentralisation and local uniqueness', *Regional Studies*, 17: 73–89.

Massey, D. (1984) *Spatial Divisions of Labour: Social Structures and the Geography of Prediction.* London: Macmillan.

Massey, D. (1993) 'Politics and space/time', in Keith, M. and Pile, S. (eds), *Place and the Politics of Identity.* London: Routledge. pp. 141–61.

Massey, D. (1996) 'Masculinity, dualisms and high technology', in Duncan, N. (ed.), *BodySpace.* London: Routledge. pp. 109–26.

Massey, D. (1999) 'Spaces of politics', in Massey, D., Allen, J. and Sarre, P. (eds), *Human Geography Today.* Cambridge: Polity Press. pp. 279–94.

Massey, D. (2000) 'Entanglements of power: reflections', in Sharp, J.P., Routledge, P., Philo, C. and Paddison, R. (eds), *Entanglements of Power: Geographies of Domination/ Resistance.* London: Routledge. pp. 269–78.

Mattingly, D. (2001) 'Place, teenagers and representations: lessons from a community theatre project', *Social and Cultural Geography*, 2: 445–59.

May, J. (1996) 'Globalisation and the politics of place: place and identity in an inner London neighbourhood', *Transactions of the Institute of British Geographers*, NS, 21: 194–215.

Melucci, A. (1996) *Challenging Codes: Collective Action in the Information Age.* Cambridge: Cambridge University Press.

Merrifield, A. (2000) 'The general law of US capitalist accumulation: contingent work and the working class', *Antipode*, 32: 176–98.

Metge, J. (1967) *The Maoris of New Zealand.* London: Routledge and Kegan Paul.

Miles, R. (1989) *Racism.* London: Tavistock.

Mills, S. (1996) 'Gender and colonial space', *Gender, Place and Culture*, 3: 125–47.

Milroy, B. and Wismer, S. (1994) 'Communities, work and public/private sphere models', *Gender, Place and Culture*, 1: 71–90.

Momsen, J. (1980) 'Women in Canadian geography', *Canadian Geography*, 24: 177–83.

Monk, J. and Hanson, S. (1982) 'On not excluding half the human from human geography', *Professional Geographer*, 34: 11–23.

Monk, J. and Liepins, R. (2000) 'Writing on/across the margins', *Australian Geographical Studies*, 38: 344–51.

Moore, D.S. (1997) 'Remapping resistance: "ground for struggle" and the politics of place', in Pile, S. and Keith, M. (eds), *Geographies of Resistance.* London: Routledge. pp. 87–106.

Moos, A. and Dear, M. (1996) 'Structuration theory in urban analysis 1: theoretical exegesis', *Environment and Planning A*, 18: 231–52.

Morin, K.M. (1995) 'A "female Columbus" in 1887 America: marking new social territory', *Gender, Place and Culture*, 2: 191–208.

Morin, K.M. (1996) Gender, Imperialism, and the Western American Landscapes of Victorian Women Travellers, 1874–1897. Unpublished PhD thesis, Department of Geography, University of Nebraska, Lincoln, NB.

Morin, K.M. (1998a) 'Trains through the plains: the Great Plains landscape of Victorian women travellers', *Great Plains Quarterly*, 18: 235–56.

Morin, K.M. (1998b) 'British women travellers and constructions of racial differences across the 19th century American West', *Transactions of the Institute of British Geographers*, 23: 311–30.

Morin, K.M. (1999) 'Peak practices: Englishwomen's "heroic" adventures in the 19th century American West', *Annals of the Association of American Geographers*, 89: 489–514.

Morin, K.M. (2002) 'Post colonialism and Native American geographies: the letters of Rosalie La Flesche Farley, 1896–1899', *Cultural Geographies*, 9: 158–80.

Morin, K.M. (2003) 'Mining empire: journalists in the American West, ca 1870', in Blunt, A. and McEwan, C. (eds), *Postcolonial Geographies*. New York and London: Continuum. pp. 152–68.

Morin, K.M., Longhurst, R. and Johnston, L. (2001) '(Troubling) spaces of mountains and men: New Zealand's Mount Cook and Hermitage Lodge', *Social and Cultural Geography*, 2: 117–39.

Morrill, R.L. (1965) 'The Negro ghetto: problems and alternatives', *The Geographical Review*, 55: 339–61.

Morrill, R.L. (1968) 'Waves of spatial diffusion', *Journal of Regional Science*, 8: 1–18.

Morrill, R.L. (1993) 'Classics in human geography revisited: author's response', *Progress in Human Geography*, 17: 352–3.

Morris, C. and Evans, N. (2001) '"Cheese makers are always women": gendered representations of farm life in the agricultural press', *Gender Place and Culture*, 8: 375–90.

Moss, P., Eyles, J., Dyck, L. and Rose, D. (1993) 'Focus: feminism as method', *The Canadian Geographer*, 37: 48–61.

Mowl, G., Pain, R. and Talbot, C. (2000) 'The ageing body and the homespace', *Area*, 32: 189–98.

Munt, S. (1995) 'The lesbian *flâneur*', in Bell, D. and Valentine, G. (eds), *Mapping Desire: Geographies of Sexualities*. London and New York: Routledge. pp. 114–25.

Murdoch, J. (1997) 'Towards a geography of heterogeneous associations', *Progress in Human Geography*, 21: 321–37.

Murdoch, J. and Marsden, T. (1994) *Reconstituting Rurality: Class, Community and Power in the Development Process*. London: UCL Press.

Murdoch, J. and Marsden, T. (1995) 'The spatialization of politics: local and national actor–spaces in environmental conflict', *Transactions of the Institute of British Geographers*, 20: 368–80.

Murray, A. (1995) 'Femme on the streets, butch in the sheets (a play on whores)', in Bell, D. and Valentine, G. (eds), *Mapping Desire: Geographies of Sexualities*. London and New York: Routledge. pp. 66–74.

Myslik, W.D. (1996) 'Renegotiating the social/sexual identities of places: gay communities as safe havens or sites of resistance?', in Duncan, N. (ed.), *Bodyspace: Destabilizing Geographies of Gender and Sexuality*. London: Routledge. pp. 156–69.

Nairn, K. (1999) 'Embodied fieldwork', *Journal of Geography*, 98: 272–82.

Nash, C. (1996) 'Men again: Irish masculinity, nature and nationhood in the early twentieth century', *Ecumene*, 3: 427–52.

Nash, C. (2000) 'Performativity in practice: some recent work in cultural geography', *Progress in Human Geography*, 25: 653–64.

Natter, W. and Jones, J.P. (1997) 'Identity, space and other uncertainties', in Benko, G. and Strohmayer, U. (eds), *Space and Social Theory: Interpreting Modernity and Postmodernity*. Oxford: Blackwell. pp. 141–61.

Newby, H. (1977) *The Deferential Worker: A Study of Farm Workers in East Anglia*. London: Allen Lane.

Nicolson, F. (1994) 'Building confidence to influence the future', *Australian Farm Journal*, March. pp. 37–8.

Oberschall, A. (1973) *Social Conflicts and Social Movements*. Englewood Cliffs, NJ: Prentice-Hall.

Paddison, R. (1983) *The Fragmented State: The Political Geography of Power*. Oxford: Blackwell.

Page, C. (2000) 'Police targeting homosexual activity in garden', *Otago Daily Times*, 21 December. p. 1.

Pain, R. (1991) 'Space, sexual violence and social control: integrating geographical and feminist analyses of women's fear of crime', *Progress in Human Geography*, 15: 415–31.

Pain, R. (forthcoming) 'Social geography, relevance and action', *Progress in Human Geography*.

Pain, R., Barke, M., Fuller, D., Gough, J., MacFarlane, R. and Mowl, G. (2001) *Introducing Social Geographies*. London: Arnold.

Painter, J. (1995) *Politics, Geography and 'Political Geography': A Critical Perspective*. London: Arnold.

Panelli, R. (2001) 'Narratives of community and change in a contemporary rural setting: the case of Duaringa, Queensland', *Australian Geographical Studies*, 39: 156–66.

Panelli, R. (2002a) 'Contradictory identities and political choices: "Women in Agriculture" in Australia', in Yeoh, B.S.A., Teo, P. and Huang, S. (eds), *Gender Politics in the Asia-Pacific Region*. London: Routledge. pp. 137–55.

Panelli, R. (2002b) 'Special issue: young rural lives', *Journal of Rural Studies*, 18, issue 2.

Panelli, R. and Gallagher, L.M. (2003) '"It's your whole way of life really": negotiating work, health and gender', *Health and Place*, 9: 95–105.

Panelli, R., Nairn, K. and McCormack, J. (2002) '"We make our own fun": reading the politics of youth with(in) community', *Sociologia Ruralis*, 42: 15–39.

Park, R.U., Burges, E.W. and McKenzie, R.D. (1925) *The City*. Chicago: University of Chicago Press.

Pawson, E. (1992) 'Two New Zealands: Māori and European', in Anderson, K. and Gayle, F. (eds), *Inventing Places Studies in Cultural Geography*. London: Longman Cheshire. pp. 15–33.

Peach, C. (1993) 'Commentary 2', *Progress in Human Geography*, 17: 350–2.

Peake, L. (1993) '"Race" and sexuality: challenging the patriarchal structuring of urban social space', *Environment and Planning D: Society and Space*, 11: 415–32.

Peet, R. (1992) 'Some critical questions for anti-essentialism', *Antipode*, 24: 113–30.

Peet, R.J. and Lyons, J.V. (1981) 'Marxism: dialectical materialism, social formations and geographic relations', in Harvey, M.E. and Holly, B.P. (eds), *Themes in Geographic Thought*. London: Croom Helm. pp. 187–206.

Penrose, J. and Jackson, P. (1994) 'Conclusion: identity and the politics of difference', in Jackson, P. and Penrose, J. (eds), *Constructions of Race, Place and Nation.* Minneapolis, MN: Minnesota University Press. pp. 202–9.

Phillips, M. (1993) 'Rural gentrification and the processes of class colonization', *Journal of Rural Studies*, 9: 123–40.

Phillips, R. (1997) *Mapping Men and Empire: A Geography of Adventure.* London: Routledge.

Philo, C. (1992) 'Foucault's geography', *Environment and Planning D: Society and Space*, 10: 137–61.

Pickles, J. (1985) *Phenomenology, Science and Geography: Spatiality and the Human Sciences.* Cambridge: Cambridge University Press.

Pile, S. and Keith, M. (eds) (1997) *Geographies of Resistance.* London: Routledge.

Pile, S. and Thrift, N. (eds) (1995) *Mapping the Subject: Geographies of Cultural Transformation.* London and New York: Routledge.

Podmore, J.A. (2001) 'Lesbians in the crowd: gender, sexuality and visibility along Montreal's Boul. St Laurent', *Gender Place and Culture*, 8: 333–55.

Pratt, G. (1982) 'Class analysis and urban domestic property: a critical re-examination', *International Journal of Urban and Regional Research*, 6: 481–502.

Pratt, G. (1986a) 'Housing consumption sectors and political response in urban Canada', *Environment and Planning D: Society and Space*, 4: 165–82.

Pratt, G. (1986b) 'Housing tenure and social cleavages in urban Canada', *Annals of the Association of American Geographers*, 76: 366–80.

Pratt, G. (1989) 'Reproduction, class and spatial structure of the city', in Thrift, N. and Peet, R. (eds), *New Models in Geography.* London and Winchester, MA: Unwin Hyman. pp. 84–108.

Pratt, G. (1992) 'Spatial metaphors and speaking positions', *Environment and Planning D: Society and Space*, 10: 241–3.

Pratt, G. (1993) 'Reflections on poststructuralism and feminist empirics, theory and practice', *Antipode*, 25: 51–63.

Pratt, G. (1997) 'Stereotypes and ambivalence: the social construction of domestic workers in Vancouver, BC', *Gender Place and Culture*, 4: 159–77.

Pratt, G. (1999a) 'From registered nurses to registered nanny: discursive geographies of Filipina domestic workers in Vancouver, BC', *Economic Geography*, 75: 215–36.

Pratt, G. (1999b) 'Geographies of identity and difference: marking boundaries', in Massey, D., Allen, J. and Sarre, P. (eds), *Human Geography Today.* Cambridge: Polity Press. pp. 151–67.

Pratt, G. (2000a) 'Gender and geography', in Johnston, R.J., Gregory, D., Pratt, G. and Watts, M. (eds), *Dictionary of Human Geography.* Oxford: Blackwell. pp. 290–1.

Pratt, G. (2000b) 'Research performances', *Environment and Planning D: Society and Space*, 18: 639–51.

Pratt, G. (2002) 'Collaborating across our differences', *Gender, Place and Culture*, 9: 195–200.

Pratt, G. and Hanson, S. (1988) 'Spatial dimensions of the gender division of labour in a local labour market', *Urban Geography*, 64: 299–321.

Pratt, G. and Hanson, S. (1991) 'On the links between home and work: family strategies in a buoyant labour market', *International Journal of Urban and Regional Research*, 15: 55–74.

Pratt, G. and Hanson, S. (1994) 'Geography and construction of difference', *Gender, Place and Culture*, 1: 5–29.

Proctor, J. and Smith, D. (eds) (1999) *Geography and Ethics: Journeys through a Moral Terrain*. London: Routledge.

The Professional Geographer (1994) 'Women in the field: critical feminist methodologies and theoretical perspectives', *The Professional Geographer*, 46(1).

Pulvirenti, M. (2000) 'The morality of immigrant home ownership: gender, work and Italian–Australian sistemazione', *Australian Geographer*, 31: 237–49.

Radcliffe, S. (1999) 'Popular and state discourses of power', in Massey, D., Allen, J. and Sarre, P. (eds), *Human Geography Today*. Cambridge: Polity Press. pp. 219–42.

Radcliffe, S. and Westwood, S. (eds) (1993) *'Viva': Women and Popular Protest in Latin America*. London and New York: Routledge.

Reed, A.W. (2001) *The Reed Concise Māori Dictionary: Te Papakupu Rāpopoto a Reed*, 6th edn. Revised by T. Kāretu. Auckland: Reed Publishing.

Rees, P.H. (1971) 'Factorial ecology: an extended definition, survey, and critique of the field', *Economic Geography*, 47: 220–33.

Rees, P., Phillips, D. and Medway, D. (1995) 'The socio-economic geography of ethnic groups in two northern British cities', *Environment and Planning A*, 27: 557–91.

Relph, E. (1970) 'An inquiry into the relations between phenomenology and geography', *Canadian Geographer*, 14: 193–201.

Resnick, S. and Wolff, R. (1987) *Knowledge and Class*. Chicago: University of Chicago Press.

Roberts, R. (1995) 'A "fair go for all"? Discrimination and the experiences of some men who have sex with men in the bush', in Share, P. (ed.), *Communication and Culture in Rural Areas*. Wagga Wagga: Centre for Rural Social Research, Charles Sturt University. pp. 151–74.

Roberts, S. (2000) 'Realizing critical geographies of the university', *Antipode*, 32: 230–44.

Robinson, V. (1981) 'Segregation and simulation: a re-evaluation and case study', in Jackson, P. and Smith, S. (eds), *Social Interaction and Ethnic Segregation*. London: Academic Press for Institute of British Geographers.

Robinson, V. (1993) '"Race", gender, and internal migration within England and Wales', *Environment and Planning A*, 25: 1453–65.

Robson, B.T. (1975) *Urban Social Areas*. London: Oxford University Press.

Rose, D. (1987) 'Home ownership, subsistence and historical change: the mining district of West Cornwall in the late nineteenth century', in Thrift, N. and Williams, P. (eds), *Class and Space: The Making of Urban Society*. London: Routledge and Kegan Paul. pp. 108–53.

Rose, G. (1993) *Feminism and Geography: The Limits of Geographical Knowledge*. London: Polity Press.

Rose, G. (1997) 'Situated knowledge: positionality, reflexivities and other tactics', *Progress in Human Geography*, 21: 305–20.

Rose, G. (1999) 'Performing space', in Massey, D., Allen, J. and Sarre, P. (eds), *Human Geography Today*. Cambridge: Polity Press. pp. 247–59.

Rose, G. and Thrift, N. (2000) 'Special issues: spaces of performance – parts 1 and 2', *Environment and Planning D: Society and Space*, 18, issues 4 and 5.

Rose, H.M. (1993) 'Commentary 1', *Progress in Human Geography*, 17: 349–50.

Rothenberg, T. (1995) '"And she told two friends": lesbians creating urban social space', in Bell, D. and Valentine, G. (eds), *Mapping Desire: Geographies of Sexualities*. London and New York: Routledge. pp. 165–81.

Routledge, P. (1993) *Terrains of Resistance: Non-violent Social Movements and the Contestation of Place in India*. Westport, CT: Praeger.

Routledge, P. (1996) 'Critical geopolitics and terrains of resistance', *Political Geography*, 15: 509–31.

Routledge, P. (1997a) 'A spatiality of resistances: theory and practice in Nepal's revolution of 1990', in Pile, S. and Keith, M. (eds), *Geographies of Resistance*. London: Routledge. pp. 68–86.

Routledge, P. (1997b) 'The imagineering of resistance: Pollock Free State and the practice of postmodern politics', *Transactions of the Institute of British Geographers*, 22: 359–76.

Russell, J.A. and Pratt, G. (1980) 'A description of the affective quality attributed to environments', *Journal of Personality and Social Psychology*, 38: 311–22.

Said, E.W. (1978) *Orientalism*. New York: Pantheon.

Samuels, M. (1978) 'Existentialism and human geography', in Ley, D. and Samuels, S. (eds), *Humanistic Geography: Prospects and Problems*. Chicago: Maaroufa Press. pp. 22–40.

Saunders, P. (1984) 'Beyond housing classes: the sociological significance of private property rights in the means of consumption', *International Journal of Urban and Regional Research*, 8: 201–27.

Saunders, P. (1990) *Social Class and Stratification*. London: Routledge.

Schaefer, F.K. (1953) 'Exceptionalism in geography: a methodological examination', *Annals of the Association of American Geographers*, 43: 226–49.

Schatzki, T. (1991) 'Spatial ontology and explanation', *Annals of the Association of American Geographers*, 81: 650–70.

Schoenberger, E. (1988) 'Multinational corporations and the new international division of labour: a critical appraisal', *International Regional Science Review*, 11: 105–19.

Seamon, D. (1980a) 'Afterword: community, place and environment', in Buttimer, A. and Seamon, D. (eds), *The Human Experience of Space and Place*. London: Croom Helm. pp. 188–96.

Seamon, D. (1980b) 'Body-subject, time–space routines and place–ballets', in Buttimer, A. and Seamon, D. (eds), *The Human Experience of Space and Place*. London: Croom Helm. pp. 148–65.

Sharp, J.P. (1996) 'Gendering nationhood: a feminist engagement with national identity', in Duncan, N. (ed.), *BodySpace: Destabilizing Geographies of Gender and Sexuality*. London: Routledge. pp. 97–108.

Sharp, J.P. (2000) *Condensing The Cold War: Reader's Digest and American Identity* Minneapolis, MN: University of Minnesota Press.

Sharp, J.P., Routledge, P., Philo, C. and Paddison, R. (2000) 'Entanglements of power: geographies of domination/resistance', in Sharp, J.P., Routledge, P., Philo, C. and Paddison, R. (eds), *Entanglements of Power: Geographies of Domination/Resistance*. London: Routledge. pp. 1–42.

Shotter, J. (1993) *Cultural Politics of Everyday Life*. Toronto: University of Toronto Press.

Sibley, D. (1995) *Geographies of Exclusion: Society and Difference in the West*. London: Routledge.

Skelton, T. (1995) '"Boom, bye, bye": Jamaican raga and gay resistance', in Bell, D. and Valentine, G. (eds), *Mapping Desire: Geographies of Sexualities*. London and New York: Routledge. pp. 264–83.

Smith, D.M. (1977) *Human Geography: A Welfare Approach*. London: Edward Arnold.

Smith, D.M. (1979) *Where the Grass is Greener: Geographical Perspectives on Inequalities*. London: Croom Helm.

Smith, D.M. (1987) 'Knowing your place: class, politics and ethnicity in Chicago and Birmingham, 1890–1983', in Thrift, N. and Williams, P. (eds), *Class and Space: The Making of Urban Society*. London: Routledge and Kegan Paul. pp. 276–305.

Smith, D.M. (1999) 'Geography, community, and morality', *Environment and Planning A*, 31: 19–35.

Smith, G. (1985) 'Ethnic nationalism in the Soviet Union: territory, cleavage and control', *Environment and Planning C: Government and Policy*, 3: 49–73.

Smith, G.D. and Winchester, H.P.M. (1998) 'Negotiating space: alternative masculinities at the work/home boundary', *Australian Geographer*, 29: 327–39.

Smith, L.T. (1999) *Decolonizing Methodologies: Research and Indigenous Peoples*. Dunedin: Otago University Press.

Smith, N. (1979) 'Gentrification and capital: practice and ideology in Society Hill', *Antipode*, 11: 24–35.

Smith, N. (1984) *Uneven Development: Nature, Capital and the Production of Space*. Oxford: Basil Blackwell.

Smith, N. (1992a) 'Gentrification and uneven development', *Economic Geography*, 58: 139–55.

Smith, N. (1992b) 'Geography, difference and the politics of scale', in Doherty, J., Graham, E. and Malek, M. (eds), *Postmodernism and the Social Sciences*. London: Macmillan. pp. 57–79.

Smith, N. (1993) 'Homeless/global: scaling places', in Bird, J., Curtis, B., Putnam, T. and Ticker, L. (eds), *Mapping the Futures: Local Cultures, Global Change*. London and New York: Routledge. pp. 87–119.

Smith, N. (1996) *The New Urban Frontier: Gentrification and the Revanchist City*. New York: Routledge.

Smith, N. (2000) 'What happened to class?', *Environment and Planning A*, 32: 1011–32.

Smith, P. (1988) *Discerning the Subject*. Minneapolis, MN: University of Minnesota Press.

Smith, S.J. (1987) 'Fear of crime: beyond a geography of deviance', *Progress in Human Geography*, 11: 1–23.

Soja, E. (1996) *Thirdspace: Journeys to Los Angeles and other Real-and-Imagined Places*. Cambridge, MA: Blackwell.

Soja, E. (1999) 'Thirdspace: expanding the scope of the geographical imagination', in Massey, D., Allen, J. and Sarre, P. (eds), *Human Geography Today*. Cambridge: Polity Press. pp. 260–78.

Soja, E. and Hooper, B. (1993) 'The spaces that difference makes', in Keith, M. and Pile, S. (eds), *Place and the Politics of Identity*. London: Routledge. pp. 183–205.

Somers, M. (1994) 'The narrative constitution of identity: a relational and network approach', *Theory and Society*, 23: 605–49.

Spivak, G.C. (1990) *The Post-Colonial Critic: Interviews, Strategies, Dialogues*. London: Routledge.

Spivak, G.C. (1996) 'Subaltern talk', in Landry, D. and Maclean, G. (eds), *The Spivak Reader*. London: Routledge. pp. 287–308.

Spoonley, P. (1994) 'Racism and ethnicity', in Spoonley, P., Pearson, D. and Shirley, L. (eds), *New Zealand Society*. Palmerston North: Dunmore Press. pp. 82–97.

Staheli, L. and Thompson, A. (1997) 'Citizenship, community and struggles for public space', *Professional Geographer*, 49: 28–38.

Stebbing, S. (1984) 'Women's roles and rural society', in Bradly, T. and Lowe, P. (eds), *Locality and Rurality*. Norwich: Geobooks.

Stewart, A. (1850) *A Compendium of Modern Geography*. Edinburgh: Oliver and Boyd.

Stokes, E. (1987) 'Māori geography or geography of Māoris', *New Zealand Geographer*, 43: 118–23.

Symanski, R. (1974) 'Prostitution in Nevada', *Annals of the Association of American Geographers*, 64: 357–77.

Symanski, R. (1981) *The Immoral Landscape: Female Prostitution in Western Societies*. Toronto: Butterworths.

Teariki, C. (1992) 'Ethical issues in research from a Māori perspective', *New Zealand Geographer*, 48: 85–6.

Teariki, C., Spoonley, P. and Tomoana, N. (1992) *The Politics and Process of Research for Māori*. Palmerston North: Massey University.

Tervo, M. (2001) 'Nationalism, sports and gender in Finnish sports journalism in the early twentieth century', *Gender, Place and Culture*, 8: 357–73.

Thrift, N. (1987) 'The geography of late twentieth-century class formation', in Thrift, N. and Williams, P. (eds), *Class and Space: The Making of Urban Society*. London: Routledge and Kegan Paul. pp. 207–53.

Thrift, N. (1996) *Spatial Formations*. London: Sage.

Thrift, N. (1997) 'The still point: resistance, expressive embodiment and dance', in Pile, S. and Keith, M. (eds), *Geographies of Resistance*. London: Routledge. pp. 124–51.

Thrift, N. (2000) 'Entanglements of power: shadows?', in Sharp, J.P., Routledge, P., Philo, C. and Paddison, R. (eds), *Entanglements of Power: Geographies of Domination/Resistance*. London: Routledge. pp. 269–78.

Thrift, N. and Williams, P. (eds) (1987) *Class and Space: The Making of Urban Society*. London: Routledge and Kegan Paul.

Tickell, A. (1995) 'Reflections on "Activism and the academy"', *Environment and Planning D: Society and Space*, 13: 235–7.

Tivers, J. (1978) 'How the other half lives: the geographical study of women', *Area*, 4: 301–6.

Tivers, J. (1985) *Women Attached: The Daily Lives of Women with Young Children*. London: Croom Helm.

Tong, R. (1992) *Feminist Thought: A Comprehensive Introduction*. London: Routledge.

Tonkin, L. (2000) 'Women of steel: constructing and contesting new gendered geographies of work in the Australian steel industry', *Antipode*, 32: 115–34.

Tuan, Y.-F. (1976) 'Humanistic geography', *Annals of the Association of American Geographers*, 74: 353–74.

Unwin, T. (1992) *The Place of Geography*. London: Longman.

Urry, J. (1987) 'The growth of scientific management: transformations in class structure and class struggle', in Thrift, N. and Williams, P. (eds), *Class and Space: The Making of Urban Society*. London: Routledge and Kegan Paul. pp. 254–75.

Urry, J. (1995a) 'A middle-class countryside?', in Butler, T. and Savage, M. (eds), *Social Change and the Middle Classes*. London: UCL Press. pp. 205–19.

Urry, J. (1995b) *Consuming Places*. London: Routledge.

Valentine, G. (1989) 'The geography of women's fear', *Area*, 21: 385–90.

Valentine, G. (1990) 'Women's fear and the design of public space', *Built Environment*, 16: 288–303.

Valentine, G. (1992) 'Images of danger: women's sources of information about the spatial distribution of male violence', *Area*, 24: 22–9.

Valentine, G. (1993a) '(Hetero)sexing space: Lesbian perceptions and experiences of everyday spaces', *Environment and Planning D: Society and Space*, 11: 395–413.

Valentine, G. (1993b) 'Negotiating and managing multiple sexual identities: lesbian time–space strategies', *Transactions of the Institute of British Geographers*, NS, 18: 237–48.

Valentine, G. (1997) 'Lesbian separatist communities in the United States', in Cloke, P. and Little, J. (eds), *Contested Countryside Cultures: Otherness, Marginalisation and Rurality*. London: Routledge. pp. 109–22.

Valentine, G. (1998) '"Sticks and stones may break my bones": a personal geography of harassment', *Antipode*, 30: 305–32.

Valentine, G. (1999) 'Eating in: home, consumption and identity', *The Sociological Review*, 47: 491–524.

Valentine, G. (2000) 'Exploring children and young people's narratives of identity', *Geoforum*, 31: 257–67.

Valentine, G. (2001) *Social Geographies: Space and Society*. Harlow: Prentice-Hall.

Valentine, G. and Holloway, S.L. (2002) 'People, place, and region cyber kids? Exploring children's identities and social networks in on-line and off-line worlds', *Annals of the Association of American Geographers*, 92: 302–19.

Valins, O.A. (1999) Identity, Space and boundaries: Ultra-Orthodox Judaism in Contemporary Britain. Unpublished PhD thesis, Department of Geography and Topographic Science, University of Glasgow, Glasgow.

Waitt, G., McGuirk, P., Dunn, K., Hartig, K. and Burley, I. (2000) *Introducing Human Geography: Globalisation, Difference and Inequality*. Frenchs Forest, NSW, Australia: Pearson.

Walby, S. (1990) *Theorising Patriarchy*. Oxford: Blackwell.

Walker, R. (1985) 'Class, division of labour and employment in space', in Gregory, D. and Urry, J. (eds), *Social Relations and Spatial Structures*. London: Macmillan. pp. 164–89.

Wall, M. (2000) 'The popular and geography: music and racialized identities in Aotearoa/New Zealand', in Cook, I., Crouch, D., Naylor, S. and Ryan, J. (eds), *Cultural Turns/Geographical Turns: Perspectives on Cultural Geography*. Harlow: Prentice-Hall. pp. 75–87.

Watson, S. (1988) *Accommodating Inequality: Gender and Housing*. Sydney: Allen and Unwin.

Watt, P. (1998) 'Going out of town: youth, "race", and place in the South East of England', *Environment and Planning D: Society and Space*, 16: 687–703.

Weber, M. (1961) *From Max Weber: Essays in Sociology.* London: Routledge and Kegan Paul.

Weber, M. (1976/1930) *The Protestant Ethic and the Spirit of Capitalism.* London: George Allen and Unwin.

Weedon, C. (1987) *Feminist Practice and Poststructuralist Theory.* Oxford and Cambridge, MA: Blackwell.

Weightman, B. (1981) 'Commentary: towards a geography of the gay community', *Journal of Cultural Geography*, 1: 106–12.

Wekerle, T. (1984) 'A woman's place is in the city', *Antipode*, 6: 11–19.

Whatmore, S. (1991) *Farming Women: Gender, Work and Family Enterprise.* London: Macmillan.

Whatmore, S. (1999) 'Hybrid geographies: rethinking the human in human geography', in Massey, D., Allen, J. and Sarre, P. (eds), *Human Geography Today.* Cambridge: Polity Press. pp. 22–39.

Whatmore, S., Munton, R., Marsden, T. and Little, J. (1986) 'Towards a political economy of capitalist agriculture: a British perspective', *International Journal of Urban and Regional Research*, 4: 498–521.

Whatmore, S., Munton, R., Marsden, T. and Little, J. (1987a) 'Interpreting a relational typology of farm businesses in southern England', *Sociologia Ruralis*, 27: 103–22.

Whatmore, S., Munton, R., Little, J. and Marsden, T. (1987b) 'Towards a typology of farm businesses in southern England', *Sociologia Ruralis*, 27: 21–37.

Whitehead, S.M. (2002) *Men and Masculinities.* Cambridge: Polity Press.

Widdowfield, R. (2000) 'The place of emotions in academic research', *Area*, 32: 199–208.

Wills, J. (2001) 'Community unionism and trade union renewal in the UK: moving beyond the fragments at last?', *Transactions of the Institute of British Geographers*, 26: 465–83.

Wincapaw, C. (2000) 'The virtual spaces of lesbian and bisexual women's electronic mailing lists', in Valentine, G. (ed.), *From Nowhere to Everywhere: Lesbian Geographies.* New York: Harrington Park Press. pp. 45–59.

Winchester, H.P.M. and Costello, L.N. (1995) 'Living on the street: social organisation and gender relations of Australian street kids', *Environment and Planning D: Society and Space*, 13: 329–48.

Winkler, J.A. (2000) 'Faculty reappointment, tenure, and promotion: barriers for women', *The Professional Geographer*, 52: 737–50.

Women and Geography Study Group of the Institute of British Geographers (1984) *Geography and Gender.* London: Explorations in Feminism Collective in association with Hutchinson.

Woods, M.J. (1997) 'Discourses of power and rurality: local politics in Somerset in the 20th century', *Political Geography*, 16: 453–78.

Woods, M.J. (1998a) 'Advocating rurality: the repositioning of rural local government', *Journal of Rural Studies*, 14: 13–26.

Woods, M.J. (1998b) 'Researching rural conflicts: hunting, local politics and actor networks', *Journal of Rural Studies*, 14: 321–40.

Woods, M.J. (1998c) 'Rethinking elites: networks, space and local politics', *Environment and Planning A*, 30: 2101–20.

Woods, M.J. (2003) 'Deconstructing rural protest: the emergence of a new social movement', *Journal of Rural Studies*, 19: 309–25.

Woods, M.J. (forthcoming) 'Political articulation: the modalities of new critical politics of citizenship', in Cloke, P., Marsden, T. and Mooney, P. (eds), *Handbook of Rural Studies*. London: Sage.

Woods, M.J., Edwards, B., Goodwin, M. and Pemberton, S. (2001) 'Partnerships, power and scale in rural governance', *Environment and Planning C: Government and Policy*, 19: 289–310.

Woods, R.I. (1970) *The Stochastic Analysis of Immigrant Distributions*. Oxford: School of Geography Research Papers, No. 11.

Woodward, R. (2000) 'Warrior heroes and little green men: soldiers, military training and the construction of rural masculinities', *Rural Sociology*, 65: 640–57.

Wright, E.O. (1979) *Class Structure and Income Determination*. New York: Academic Press.

Wright, E.O. (1985) *Classes*. London: Verso.

Zavella, P. (2000) 'Latinos in the USA: Changing socio-economic patterns', *Social and Cultural Geography*, 1: 155–67.

Zelinsky, W., Monk, J. and Hanson, S. (1982) 'Women and geography: a review and a prospectus', *Progress in Human Geography*, 6: 357–66.

Index